Forces of Nature

The Women Who Changed Science

↓

To all the lady scientists and doctors, computers, wives, sisters, domestic engineers, and popularizers who have always been at the heart of scientific inquiry and discovery.

Forces of Nature

The Women Who Changed Science

Anna Reser & Leila McNeill

F FRANCES
F LINCOLN

Contents

Section IV
The Twentieth Century, Pre-World War II

Section V
The Twentieth Century, Post-World War II

Afterword

Introduction

Reading Women's Silence in
the History of Science

Writing the history of women is often an encounter with absence and silence. In the history of science in particular, the written record simply does not reflect the number of women who have always participated in science and medicine, nor does it reflect the complexity of their stories. In many times and places, it was seen as inappropriate for women to participate in public life, particularly in the Western world, so there are simply fewer records of women in general. Access to public life confers value and status and makes a person's accomplishments, and even the basic facts of their life, worth recording. Historically, women have been denied this access, and along with it a more complete record of their lives.

In the sciences, this lack of a record is compounded by the insular nature of scientific communities. Women were often not allowed into the institutional spaces that create and keep written records, such as professional societies and scientific journals. The challenge and reward of writing women's history is reading these gaps in the record for clues to the ways that women have been erased from history, which is sometimes all we can definitively say about their lives.

The farther back in time one searches, the more difficult it becomes to locate written accounts of the lives of women in science. Very few written records of any kind exist from ancient times, and the relatively low status of women in different societies has impacted the likelihood that they will be mentioned in such records. There are, however, records about a few women whose lives and work included intentional and careful study of the natural world.

The earliest textual records of any culture come from the ancient city-states of Mesopotamia in the valleys of the Tigris and Euphrates Rivers. Enheduanna (circa 2285–2250 BCE), the very first author whose name was recorded, was a poet and priestess who lived in the ancient city of Ur. She was the daughter of the king Sargon, who likely made her a high priestess to solidify his political power. As high priestess, Enheduanna's writings, preserved on clay tablets, lent her great authority. As part of her management of the temple in Ur, she oversaw the agricultural activity on the temple lands. Because Enheduanna was also responsible for organizing religious rituals throughout the year, she must have managed a complex liturgical calendar based on the phases of the moon.[1]

Archaeologists first discovered evidence of Enheduanna in the 1920s when excavating the site of Ur. There they uncovered a stone disk carved on one side

↑ Ancient astronomers studied the motions of the planets, named the stars and constellations, and predicted lunar and solar eclipses eclipses.

Introduction

with a relief of a temple scene depicting a woman in elaborate clothing, flanked by what are probably male attendants, performing a ritual on an altar set before a ziggurat (a rectangular stepped tower.)[2] An inscription on the other side of the disk identifies the central figure as the priestess of the moon god Nanna:

> "Enheduanna, true lady of Nanna, wife of Nanna,
> Daughter of Sargon, King of all, in the temple of
> Inanna [of Ur, a dais you built (and) "Dais,"
> Table of heaven (an)] you called it."[3]

Later discoveries of Enheduanna's writings, which include long poems to Nanna, and to her personal goddess Inanna, and more than forty Temple hymns, have made her an important figure in the study of ancient Sumerian literature and culture.[4] To historians of science, the record of Enheduanna's life offers evidence of women's participation in the observation of nature even in the earliest written record of human society. Women have always known nature.

Enheduanna's life was revealed through great effort by archaeologists and linguists, working with very little material. The task of recovering women's histories becomes easier in periods when writing was more commonplace and where more of these records have survived the centuries. Perhaps the most famous scientific woman of the ancient world was Hypatia of Alexandria (circa 335–405 CE), a fourth-century Greek mathematician and philosopher who lived in the city of Alexandria nearly 3,000 years after Enheduanna. Unlike those of the famous men of Greek philosophy and science, no known copies of Hypatia's own scientific writings exist. Instead, historians have pieced together her biography from what others have written about her and contextual information about the larger world of Greek institutions of learning. Hypatia was the daughter of Theon of Alexandria, a mathematician and astronomer, and she probably received her training in those disciplines at the Museum of Alexandria, of which Theon was the head. She later became the director of the Neoplatonic school at Alexandria when she was in her thirties.[5]

Hypatia's story consists of series of textual records that extend from around the time she lived and into the nineteenth century, all of which are scattered across different types of texts, in different languages and from different cultures. It is remarkable that record of her existence survived so much translation and transference across the centuries, but it presents a challenge for historians. What conclusions can be drawn from this textual record, if not about Hypatia's own life, then about the role of women in Greek society and learned culture?

Only recently have scholars looked to Hypatia's story to uncover information about her scientific and mathematical activities. Her prominence in the written historical record was at first due to the violent circumstances of her death, which were likely entangled with Alexandrian politics.[6] Part of the problem lies

Instead of simply accepting that there are spaces where women will not be found, we should be asking why a woman might not appear in a particular sphere and who prevented her from being there.

in the assumptions that scholars have made about who participated in public life in the ancient world, and who was able or willing to engage in scientific inquiry.

Stories like that of Hypatia, whose life and work were preserved for reasons beyond her scientific endeavors, point to a much richer world of knowledge making and learning for women in the ancient period than is often assumed. It is likely that there were many women practicing astronomy, learning and teaching mathematics, and deeply engaged with philosophy, but records of their lives have simply not survived. What of the women who searched the skies and wrote about the natural world who did not suffer the same fate as Hypatia? We may never know their names, but the project of recovering the place of women in science benefits from our informed assumption that they existed and that their scientific practices, like Hypatia's, were held in the same high regard as those of their male contemporaries.

ABSENT FROM HISTORY

In the cases of Enheduanna and Hypatia, historians are fortunate to have a record of their lives and work, however fragmentary or limited they may be. But how should we go about recovering the stories of women who left no written record and whose existence may not even be known? Historians have developed methods for researching historical actors who left few written records. Sometimes, as in the case of Hypatia, a biography can be pieced together from the writings of others. This is especially true of the medieval and early modern periods where there are simply more written records of all kinds. In the case of early modern Europe, for example, scholars have found evidence of women whose lives and work intersect with the sciences by studying images and examining the catalogs of private libraries that belonged to wealthy women to see if they owned scientific books. Women were frequently in correspondence with more well-known scientists, and sometimes women are listed in the dedications of scientific texts written by men.[6] These records are important sources of information about women who may not have had the opportunity to author their own scientific works but were still very much involved in scientific life.

The history of scientific culture and the women who participated in that culture is much more well-studied in Europe than around the rest of the world. While many of the methods used to recover histories of European women in science can be adapted to study other geographic areas and time periods, it is also important that such studies take account of the diversity of scientific

↑ The disk of Enheduanna is a limestone calcite cylinder made between 2350 and 2300 BC, and shows Enheduanna and her attendants carved in relief on the front, and an inscription of Enheduanna on the reverse.

cultures. For instance, when studying science in the early modern Islamic world, historians look for scientific activity in the courts of the Islamic empire, rather than analogues to the academies and universities of Europe. Also, use of the printing press was not as widespread in the Islamic world as it was in Europe, which made the circulation of printed books significantly different in the Islamic world when compared to the scientific network in Europe.[7]

READING RECORDS AND GAPS

The contributions of women are often overlooked by history because we tend to look at the past through modern eyes. Finding women whose practices in the past resemble something we recognize as science today is quite rare. Until recent decades, historians would never have considered the practices of witches and midwives to be scientific. A particularly important method for identifying women in science involves expanding the definition of scientific activity and looking for women in spaces we might consider merely adjacent to science. In the early modern world, astrology and alchemy were both considered learned practices with similar status to astronomy or physics; it was only in later periods that these fields came to be considered fringe or pseudoscientific.

↑ An Indo-Persian celestial globe from 1362–3, showing the constellations and the ecliptic as they are seen from the surface of the earth.

Women were also active in medical practice in all periods of history. In the Islamic world, male physicians mentioned women being trained as doctors themselves, although we know of no medical books authored by women from the Islamic world exist. Often caring for their own families, some women took up positions as midwives to care for other people.[8]

Historians now recognize that these women engaged in the systematic pursuit of knowledge about nature just like male scientists and natural philosophers. We also now know that their exclusion from the institutions of knowledge-making, which we associate with scientists, had little to do with women's practices and everything to do with their gender. By looking outside a narrow understanding of what counts as science, we find evidence of women at every turn, working diligently and clamoring to be let into the hallowed halls of scientific pursuit.

Instead of simply accepting that there are spaces where women will not be found, we should be asking why a woman might not appear in a particular sphere and who prevented her from being there. Reframing the question in this way transforms the gap in the historical record into the evidence of an action, not simply something missing. In the late eighteenth and early nineteenth centuries, as science was becoming a respected profession for men, women interested in science were relegated, often intentionally, to its margins. Certain practices were deemed "appropriate" for women, such as botany, while the pathways to working in other fields, like mathematics or chemistry, were narrowed or blocked by male-dominated scientific organizations. Women turned to science writing and popularization in order to continue to feed their interest in the study of nature, while male scientists, working in the protected spaces of scientific institutions, regularly used women's scientific work without ever giving them credit. The absence of women in the collective understanding of the history of science is not always the result of a lack of evidence. Like remembering, forgetting, too, is an active practice. The contributions of women to the study of nature have been simply overlooked in many cases, a matter, where ample evidence exists, of common negligence.

Section I

Antiquity to the Middles Ages

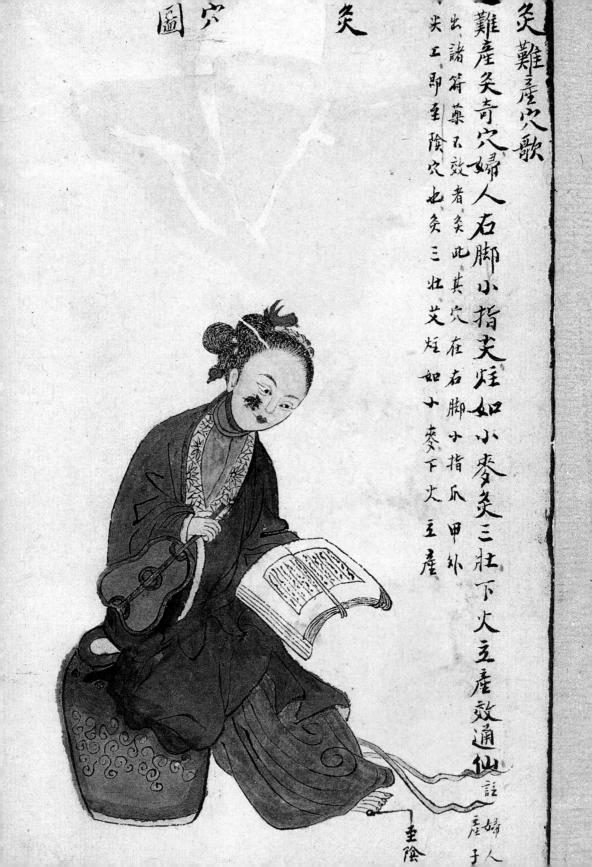

灸難產穴歌

難產灸奇穴婦人右腳小指尖炷如小麥灸三壯下火立產效通仙

註婦人產子

出諸符藥不效者灸此其穴在右腳小指爪甲外

尖上即至陰穴也灸三壯艾炷如十麥下火立產

圖穴　　灸

至陰

CHAPTER 1

Physicians, Midwives, and "Grannies"

ANCIENT BODIES AND PRACTICES

Different cultures throughout history have had very different ideas about the body, the concept of health, and the social roles of women, all of which influenced women's access to education and medical practice. We tend to think of the organization of human anatomy and the function and role of the parts of the body as a settled matter, one that was uncovered and codified long ago. But the history of medicine is the story of constant renegotiation of these bodily facts and understandings. In Greek medicine, for instance, the womb has been the defining organ separating the male body from the "imperfect" form of the female body, and the uterus and its functions have been markers of feminine nature. But in ancient Chinese medicine, in which the forces and parts of the body are organized according to a balance of masculine and feminine energy, the uterus does not have the same social role in defining sex and gender. Few records remain of women nurses, physicians, midwives, pharmacists, and surgeons from Antiquity. This fact, combined with popular narratives about the "first" woman doctors and medical practitioners, has tacitly subsumed these early women into a universal community of practitioners thereby obscuring important social and cultural contexts that variously enabled or impeded women's medical practice.

Among the earliest mentions of women working as medical practitioners in a formal sense—that is their medical practice was how they earned money or status—come from the culture of ancient Egypt. The evidence for women physicians in ancient Egypt is sparse, but it suggests that its peculiar political

← Acu-moxa point chart, showing the zhiyin (Reaching Yin) point, here also described as the "difficult birth" point, from *Chuanwu lingji lu* (*Record of Sovereign Teachings*), by Zhang Youheng, a treatise on acu-moxa in two volumes.

context may have shaped a community of women medical practitioners that was separate from that of men. Peseshet was a woman who lived in the fifth to sixth dynasty in the Old Kingdom (2465–2150 BCE), and whose stela—a funerary marker—was found inside the tomb of Akhet-hotep in Giza. Various scholars have interpreted the inscription of her title on the stela as "chief woman physician" or "lady director of lady physicians," which has led historians to believe that there may have been all-women groups of doctors, of which Peseshet was supervisor.[9] Peseshet's title suggests something of the way women participated in public life in the Old Kingdom, likely in service to other women and probably not as overseers of men. The political organization of ancient Egyptian culture excluded women from the regular bureaucracy and administration of the kingdom.[10] As a result, historians cannot offer any definitive description of the practice of women physicians since there are no records specific to it. What they can do, however, is infer about the activities of medical women by generalizing from those of men for whom we have much more evidence.

Another way to approach the thin sources on women physicians in ancient Egypt is to work from the assumption that women would have provided

↑ The Edwin Smith Papyrus, a surgical treatise from c.1600 BCE.

medical care to other women in matters of gynecology and childbirth. In these cases, we have more evidence of actual activities, which are recorded in a number of medical papyri discovered by Egyptologists in the nineteenth and early twentieth centuries, including the Ebers Papyrus and the Edwin Smith Surgical Papyrus. If we assume that some women acted as doctors by providing informal medical care to others, we can assume they performed some of the procedures detailed in these papyri. To determine if a woman was pregnant, for example, the ancient Egyptian physician could take her pulse, examine her breasts, or test her urine by sprinkling it on barley and emmer seeds to see if they germinated.[11] To prevent pregnancy, prescriptions included inserting crocodile dung or honey into the vagina.[12] The medical papyri included prescriptions for speeding childbirth or easing difficult births.[13]

WOMEN AND REPRODUCTIVE MEDICINE IN THE ANCIENT WORLD

In many cases, the only evidence we have for women's interaction with medicine—as patients or as practitioners—has to do with matters of pregnancy, childbirth, and the female reproductive system. This is certainly the case in ancient Greece and Rome. The most famous Greek writing on medicine is the Hippocratic Corpus, a group of texts on physiology, etiology, surgery, and other medical topics written by about sixty different authors (mainly in the late fifth and early fourth centuries BCE).[14] Even though Hippocrates is popularly called "the father of modern medicine," the authors of the Hippocratic Corpus were actually a group of authors who shared certain ideas about the body and disease.

The Hippocratic Corpus is especially interesting to historians because its authors shared an interest in finding rational causes and cures for disease that did not rely on magic or divine intervention. While "folk" remedies persisted in ancient Greece and Rome even after the Hippocratic Corpus, the Hippocratic writers understood such remedies in different terms. An herb that might have once been prescribed for magical reasons, for instance, was still prescribed but with an attendant rational explanation that attempted to chart the causal effects of the herb on the body.[15]

The Hippocratic Corpus contains several sections on the nature and diseases of women, from which historians have inferred the roles of women as healers and patients in the practice of medicine in the ancient Near East. Along with works by other ancient thinkers like Soranus of Ephesus, historians have constructed a fairly detailed picture of medicine in this period, including how these ideas were translated and transmitted across time and geography to form the foundation of modern Western medicine. In this last respect, as we will see throughout this book, the ideas that the Greeks constructed about the nature of women and their inferiority to men were influential and long-lasting. It is for this reason that the study of the intellectual history of medicine—how ideas about the body's functioning are created and dispersed—is an important component of women's history and the history of science.

The Greeks believed that women were inherently inferior to men, because they were in fact imperfectly formed men who suffered from a host of related ailments. Menstruation, for example, was considered an ailment. In the Hippocratic text *Diseases of Women*, the author introduced his subject by asserting "the woman who has never given birth suffers more intensely and more readily from menstruation than a woman who has given birth to a child."[16] The author offered more observations on the fundamental nature of women, noting that "a woman's flesh is more spongelike and softer than a man's," and "a woman has warmer blood than a man and therefore she is warmer than a man."[17]

The ancient Greeks also theorized that the uterus was highly mobile within the body and the womb could be displaced by not having regular intercourse or by other conditions that would cause the body to be more empty than usual, leaving room for the uterus to move.[18]

Diseases of Women contains extensive notes on therapies and prescriptions for the illnesses that befell women. In a section on diagnosing infertility, for example, the writer reminded physicians that "healing, then, involves a concern for the entire body and the healing of the cervix," and suggested certain kinds of vapor baths, purgation, and other procedures that might help the woman to conceive.

MEDICAL PRACTICE AND MIDWIFERY

While medical writings like the Hippocratic Corpus were almost certainly intended for use by men, some women in ancient Greece and Rome were midwives who would have possessed similar knowledge and therapies in their work caring for pregnant women and delivering babies. In the early second century CE, Soranus of Ephesus wrote about midwifery in his treatise on gynecology, also titled *The Diseases of Women*. Soranus placed midwifery in the context of the medical profession, detailing the ideal characteristics of a good midwife and some of her techniques.[19] The best midwives were, according to Soranus, literate and versed in medical theory; they had good memories, a neat appearance, and slender hands and fingers with trimmed nails to avoid scratching patients by accident.[20] According to historian Valerie French, "[a] number of Roman legal provisions strongly suggest that midwives enjoyed status and remuneration comparable to that of male doctors."[21]

Roman midwives carried kits of necessary supplies and also provided a special birthing chair for laboring women. The seat of the chair was crescent-shaped with an opening in the middle for the baby to pass through and into the hands of the midwife. The midwife and her assistants were also responsible for managing the emotional state of the mother by calming her anxieties, which they believed could cause a difficult birth.[22] In the event the birth did have complications, however, the midwife had numerous remedies at her disposal to speed or ease childbirth. The paws of hyenas, snake skin, and herbs like dittany were all remedial ingredients believed to aid the process of childbirth.[23]

WOMEN MEDICAL PRACTITIONERS IN ANCIENT CHINA

While the womb played a central role both in the differentiation between male and female bodies and in the so-called "diseases of women" of ancient Western medical traditions, this was not true for other parts of the world. In ancient Chinese medicine, the body was organized according to a cosmology of

↑ Ancient Roman relief carving of a midwife attending a woman giving birth. Midwifery was a recognized profession for women in ancient Rome.

balanced forces, in which the body formed a microcosm of the larger universe.[24] Similarly, family life in Imperial China was also a microcosm of the state, and many matters that may have been more publicly administered in the West took place within the family, including medicine. Historian Charlotte Furth has traced the development of *fuke* in the Song dynasty, which can be translated as gynecology but refers literally to the medicine of wives.

Within the microcosmic family of Imperial China and the Imperial court, women did practice medicine. But as in other times and places, many of the records we have of women doctors and midwives come from men's writings about them. In China, these writings were often very dismissive of women practitioners, who male doctors often saw as unlearned, crude, and wiley. Furth uses these records from the Ming period spanning 1368–1644, rather than a much smaller number of official records of women doctors, to explain the social and cultural context of women in medicine.[25] As with ancient

In many times and places, women medical practitioners have been marginalized for adherence to practices that are considered traditional or otherwise outmoded by "modern" medicine.

Egyptian medicine, records by or about women medical practitioners are very scarce. Male doctors described women medical practitioners as three of the six "grannies:" medicine sellers, shaman healers, and midwives. Stereotypes of women that a family might encounter, the grannies were among those that male writers warned families against. In the Imperial court during the Ming period, male doctors were forbidden from treating the women of the court, and thus a maid would sometimes be trained to care for the women of the court, or else a granny would be brought in.[26]

The fact that these women or "grannies" existed in numbers that warrant stereotyping indicates that they made their living as healers. These writings point to a tradition of professionalized medicine for women in this time period, whether men approved of them or not. Male medical practitioners recognized that it was necessary for women to treat other women when it was inappropriate for men to do so, but at the same time, they sought ways to regulate women's medical practices.[27] These women went from place to place, helping deliver babies, procuring and mixing medicines, and offering treatments including moxibustion and acupuncture that male physicians considered out of fashion.[28] In many times and places, women medical practitioners have been marginalized for adherence to practices that are considered traditional or otherwise outmoded by "modern" medicine.

In the ancient and medieval world, women practiced medicine in a number of formal and informal ways, with varying types of education and training. Informally as caretakers for their own families and neighbors was the most common way women practiced medicine. For people of lower social standing, access to professionalized medicine was often prohibitively expensive, but they could receive care from family members or lower-status practitioners, some of whom were women. Women were called on in particular to give medical care to other women, especially in matters of generation and childbirth. But even in the ancient period, whose reconstruction by scholars has relied on extremely sparse records, women made themselves known as healers and carers, parlaying their knowledge of medicine into a living for themselves and an essential aspect of care for their communities.

CHAPTER 2

The Supernatural and the Sanctified

On the northeastern coast of Greece, surrounded by mountains to the north, south, and west, lies ancient Thessaly. This region of Greece held particular sway over the imaginations of the ancient Greeks and Romans, and their literature is littered with references to Thessaly as a land of sorcery and magic. A possible origin for Thessaly's magical association comes from the powerful witch Medea, who is said to have flown over Thessaly in her dragon-drawn chariot and flung out a chest of herbs that landed, took root, and grew there.[29] Other stories of Medea tell of her ability to remove the moon from the night sky so that she could perform her sorcery under the cover of complete darkness. The magic of seedlings that took root in Thessaly is said to have lent power to the women there, giving rise to even more stories of Thessalian maidens who were rumored to have the power to command the moon and draw it down from the sky. Stories such as these fed long-held beliefs that women have an inherent connection to nature, both terrestrial and celestial. In Antiquity and the Middle Ages, however, when nature and especially astronomical phenomena were mysterious and unknown, women's knowledge of such unknowable things marked them for speculation and sometimes fear.

The story of Medea's herb chest and Thessaly's magical roots can be found in Aristophanes' Greek comedy *The Clouds*, written around 423 BCE. The tale was added by an anonymous scholiast (a commentator on ancient and classical texts) to give context to a dialogue between two characters in the play, Socrates and Strepsiades. The two characters have a seemingly mundane conversation

← Hildegard of Bingen envisioned the cosmos as a fiery egg, which was illuminated in her book *Scivias*, c.1175 AD.

about paying debts wherein Strepsiades says he can dodge payment by buying "a witch woman, a Thessalian, and take down the moon at night." Strepsiades jests that if he could prevent the moon from rising again, signaling the beginning of the month when debts were due, then he would not have to pay. The scholiast reported that the Thessalian women received their power from Medea's magical herbs, but who are these moon-stealing witches?

A scholia to Apollonius of Rhodes, from around the third century BCE, commented on Apollonius' famous epic poem *Argonautica*, writing that

> *"[i]n antiquity they used to think that witches drew down the moon and the sun. Accordingly, even up to the time of Democritus, many people used to call eclipses 'kathaireses.' Sosiphanes, in the Meleager says: Every Thessalian maiden with magic songs [is] a false bringer-down of the moon from the sky."*

In other words, Thessalian witches were commonly believed to have the power to cause a lunar eclipse, and Sosiphanes, another poet, believed the women to be frauds.[30] Another Greek writer named Asclepiades, who lived in the second century BCE, wrote that Thessalian women studied and learned the movements of the moon and would announce when they would bring it down. Even Plato mentions the Thessalian witches in his Socratic dialogue *Gorgias*, noting "the Thessalian enchantresses, who, as they say, bring down

↑ The so-called witch Medea bids farewell to her children born from a union with Jason, leader of the Argonauts. The lithograph, c.1880, portrays the scene from Euripides' play Medea.

the moon from heaven at the risk of their own perdition."[31] Witches drawing down the moon and removing its light from the night sky became known as the Thessalian Trick.

Despite widespread belief in the powers of Thessalian women and repeated references to their activities over the centuries, who they actually were and the particulars of their lunar ritual is shrouded in mystery. Some have speculated that they were a group of early astronomers who observed and predicted lunar eclipses, and they led people to believe that they were performing magic during the lunar phenomena. But since such events happen only about once every three years at any given place, a real lunar eclipse likely would not have provided a consistent screen for the women to perform their "trick." This has cast doubt on whether "bringing down the moon" actually had anything at all to do with eclipses, as we understand them today. Others have posited that the so-called witches performed the trick by way of elaborate theater, constructing devices with candles, mirrors, and pulleys to give an allusion of pulling down and "disappearing" the moon.[32] Whoever they were, astronomers, witches, or charlatans, their reputation was so pervasive that when lunar eclipses did occur, they and their magic were thought to be the cause.

WITCH, CHARLATAN, OR ASTRONOMER?

In the literature and records that have survived all these centuries, the name of only one potentially real Thessalian sorceress has been found. Both Greco-Roman writer Plutarch and a scholia to Apollonius identify Algaonice (circa first or second century BCE) as the daughter of Hegemon of Thessaly and as an expert astronomer. In the *Argonautica*, the scholia wrote,

> *"Aglaonice, the daughter of Hegemon, being skilled in astronomy, and knowing the eclipses of the moon, whenever it [the moon] was going to be involved with them [eclipses] used to say that she was drawing down the goddess…"*[33]

In *Conjugalia Praecepta*, an advice manual for married couples, Plutarch tells much the same story to demonstrate that a woman should be learned in geometry and astronomy so not to be deceived by Aglaonice

> *"and how she, through being thoroughly acquainted with the periods of the moon when it is subject to eclipse, and, knowing beforehand the time when the moon was due to be overtaken by the earth's shadow, imposed upon the women, and made them believe that she was drawing down the moon."*[34]

Plutarch, writing in 100 CE, mentioned Aglaonice two other times in his writings, pointing out that she was not an actual witch but a skilled astronomer who used her knowledge of natural phenomena to exploit people's ignorance.

Aglaonice did not leave behind any original texts, and some have suggested that she was nothing more than a myth like Mycale, another Thessalian witch who was said to draw down the moon "by her charms."[35] If Aglaonice did actually exist, it is unclear when exactly she lived. The manner in which Plutarch wrote about her implies that she lived before his writing about her in 100 CE, but the Greeks did not yet have much knowledge about eclipses and lunar predictions, and would not until Ptolemy's astronomical treatise *Almagest*, dated circa 150 CE. For Aglaonice to be such an expert in lunar observation, she would have needed knowledge of Babylonian astronomy from Mesopotamia. The ancient Babylonians left behind one of the largest collections and longest continual sets of astronomical records from the ancient world. Their Astronomical Diaries in particular contain datable observational texts with copious material on lunar and solar eclipses, starting around 750 BCE.[36] Babylonian astronomy did not reach Greece until around the fourth century BCE, leading some scholars to place Aglaonice around the second or first century BCE.[37] How Aglaonice would have had access to education in Babylonian astronomy is also unknown.

If the scholia and Plutarch are correct about her talents, Aglaonice must have been a skilled astronomer indeed to pull off such a scheme. What is significant about Aglaonice, whether she existed or not, is the astronomical meaning her story gave to the lunar rituals of the other Thessalian witches. Her entry into the lineage of Thessalian witchcraft has prompted a reexamination of the supposed supernatural nature of the activities of the other witches. Still, Aglaonice's power and legendary status in Antiquity came not from her knowledge of natural phenomena but her association with sorcery, ensuring that Thessaly still holds a magical place in the modern imagination.

MYSTIC VISIONS OF THE COSMOS

Even though the witches of Thessaly were sometimes called "demonic women," they lived centuries before the witch-burnings of the late Middle Ages. If they had lived then, when Christianity had crystalized the relationship between witchcraft and Satanic influence, they would have been in danger for their association with dark magic. But even without such a conduit to the occult, women with knowledge or power that fell outside of accepted social and cultural traditions have been subject to rumor and suspicion in many different time periods throughout history. This was true even for the famously devout abbess and mystic Hildegard of Bingen, who conceived of her cosmological system through divine visions and ecclesiastical exegesis.

Born in 1098 in the Rheinland Valley in what is now Germany, Hildegard (see page 33) began receiving divine visions at the age of three. During her visions, Hildegard would see intense light, which "pierced my brain, my heart

↑ "The Redeemer" from Book II of *Scivias* portrays Hildegard's vision of the creation of the heavens, earth, and humans.

and breast through and through like a flame which did not burn."[38] Through the light, she would receive divine wisdom from God, allowing her to interpret holy scripture. Through her childhood and into adulthood, Hildegard kept her visions to herself: "Out of fear of people I dared not tell anyone."[39]

In Christian mysticism, mystics believe they obtain inaccessible knowledge of God through a union with God by way of self-contemplation and insight, sometimes communicated in the form of visions. Though there were famous male mystics, such as St Augustine of Hippo and St Peter Damian, it appears that no such mystical tradition existed for women before Hildegard; she is the first known woman mystic in the West for whom we have record. As a woman, her mystical experience of the divine placed her on untrodden ground, which may explain her hesitancy to make her visions known. Within the church, women were barred from preaching according to St Paul's injunction in the New Testament in 1 Corinthians 14:33–35. Women who did speak with authority on scripture were disparaged and dismissed as hysterics and/or witches.[40] But at the age of forty-three, Hildegard began to document her visions after she heard a heavenly voice saying, "O fragile human, ashes of ashes, and filth of filth! Say and write what you see and hear."[41]

Hildegard recorded her visions in *Scivias*, an illuminated text in three parts which described twenty-six visions. In this work, Hildegard laid out her cosmological system, which in many ways conformed to other cosmological systems developed by male cosmologists of the time. Her picture of the cosmos stands apart in other ways, however. In one cosmological vision, the earth, conforming with Ptolemaic cosmology, was situated in the middle of the universe, but the universe was shaped like an egg on fire. She conceived of the earth as having four elements represented as chaos, which ensued after The Fall, and which would regain order and harmony after the Lord's judgement. The egg also contained the constellations, the moon, and Mercury and Venus. The fiery shell held Mars, Jupiter, Saturn, and the sun at the egg's apex. Some have interpreted Hildegard's egg-shaped cosmos as a specifically feminine one that privileges a maternal and nurturing image of the universe: the egg is a womb enclosing its child the earth. The cosmic pregnancy itself represented a union of matter and spirit, which begat a holy material universe.[42]

FINDING POWER IN WOMEN'S NATURAL KNOWLEDGE

In 1147, Pope Eugenius III approved the first part of *Scivias* as authentic and, subsequently, permitted Hildegard to preach in public. Hildegard then became the first woman to receive papal endorsement as an authority on theological matters.[43] Divine vision as a mystic gave Hildegard power and authority that she was otherwise denied as a woman, and her status as mystic, chosen by God himself, facilitated her ascent in both the church and society.

Hildegard's visions no doubt granted her power and cultural capital in the matters of the church and scientific discussions of the cosmos, but she remained aware of the fine line she walked as a vocal woman. Some who opposed her authority questioned why her divine wisdom should come to a woman instead of man.[44] Hildegard's authority lay only in her sanctity as one chosen by God, and a woman's sanctity was evidenced by humility.[45] Throughout her writing, she was frequently self-deprecating, calling herself "a poor little figure of a woman" as a way to display her humility before the church and mollify her detractors. When her sanctity and authority came into question, it was essential for her to show that it was God, not her, who spoke.[46]

––––––––––––

Hildegard and Aglaonice do not have much in common: one was a mystic chosen by God from the Middle Ages and one was a supposed witch trafficking in dark magic from Antiquity. What they shared was connection to the supernatural, and it is through this connection that they gained recognition in their time and ours. On its own, women's knowledge did not seem to hold much value. Women's knowledge of natural phenomena required something of the supernatural, divine or evil, to be heard and seen. And whether or not the magic was real does not much matter because it provided a real and credible screen through which women could exercise their power and authority in matters of science.

Hildegard of Bingen

(circa 1098–September 17, 1179)

Abbess, polymath, and mystic, Hildegard of Bingen was born in 1098 in
Bermersheim bei Alzey in the Rheinland Valley in what is now Germany.
Hildegard, a frail and sickly child, began receiving divine visions at the
age of three, which singled her out as strange. At the age of eight, her
parents gave over her care to the church. She was taken to an anchorage,
a small, cell-like structure, attached to Disibodenberg, a Benedictine
monastery near the Rhine River, and placed under the care of the
anchoress Jutta. Hildegard and Jutta became close, and in 1136 when Jutta
died, Hildegard was elected *magistra* (female suprerior) in Disibodenberg.

Hildegard eventually recorded her mystic visions from God in
Scivias, a beautifully illuminated three-part text which brought her fame
and papal endorsement. Her nortoriety attracted so many new people
seeking admission into the religious order that St. Disibodenberg could
not house them all. She decided to move herself and her nuns to a new
convent in Rupertsberg Monastery, twenty miles away. This monastery
became similarly overcrowded, and Hildegard proposed the opening
of a second monastery at Ebingen. When the church elders denied her
permission for this, she coincidentally fell ill, a fate she believed God had
imposed on her for delaying His will. When the elders relented, Hildegard
rose from her sickbed and opened the abbey, now the Benedictine Abbey
of St. Hildegard, in 1165.

In addition to cosmological theories, Hildegard authored two treatises
on medicine and natural history, texts on the lives of saints, and *Symphonia
armonie celestium revelationum*, a collection of original lyric poems with
an accompanying musical composition. Hildegard died on September
17, 1179 at Rupertsberg, but she was not canonized as a saint until Pope
Benedict XVI declared it in 2012. That same year, Pope Benedict also made
Hildegard a doctor of the church, making her one of only four women to
receive the honor.

Section II

The Renaissance & The Enlightenment

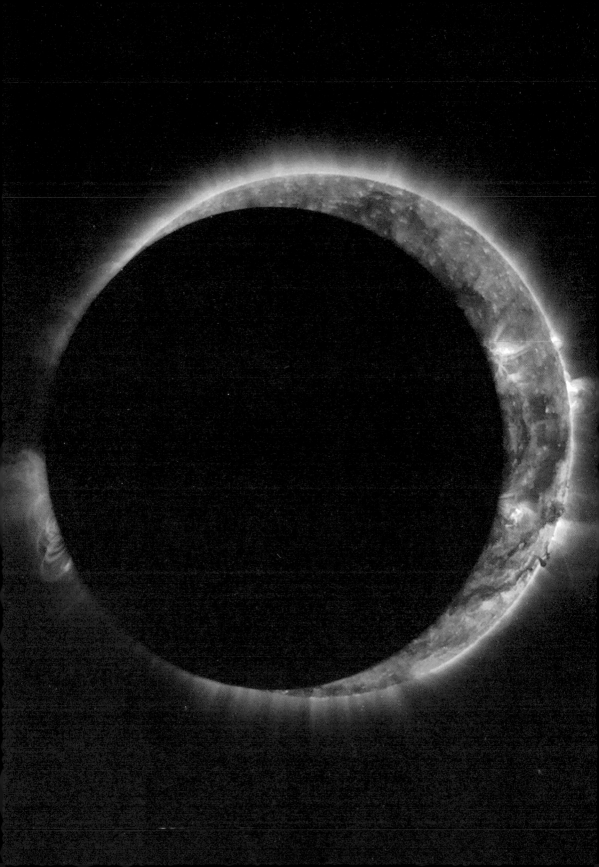

CHAPTER 3

Women Calculate Their Own Path to Science

CHARTING A PATH

In 1650, Maria Cunitz (see page 44) published *Urania Propitia,* a book of high-level mathematics and astronomical calculation of Keplerian astronomy. It remains the earliest surviving work of science written by a woman.[47] In the seventeenth century, it was rare to find women making significant contributions to science and mathematics, but not because women were unskilled. Rather, it was the result of widely accepted beliefs that women were not capable of succeeding in such endeavors, and if they were, they should not be allowed to participate. With the publication of *Urania Propitia,* Cunitz proved that women could be every bit the capable astronomers and mathematicians as their more famous male counterparts.

Cunitz was born in Central Europe in Silesia soon after Johannes Kepler published his heliocentric model of the solar system in *Astronomia Nova* (*New Astronomy*). Keplerian astronomy, which demonstrated the elliptical orbits of the planets, was one of several competing models of the solar system in Europe in the seventeenth century. Before the astronomical community had fully accepted Kepler's three laws of planetary motion (even Galileo did not readily adopt Kepler's first law of elliptical orbits),[48] Cunitz embraced his astronomical system.

Despite her adherence to Keplerian astronomy, Cunitz found errors in his 1627 *Rudolphine Tables,* a catalogue of stars, planetary tables, and directions for calculating planetary positions, and she took it upon herself to correct them. In the resulting *Urania Propitia* (*Beneficent Urania*), she corrected many of Kepler's errors and simplified his calculations by removing his logarithms. While Cunitz's work also included the kind of errors that were common for these types of tables, the accuracy of her tables exceeded Kepler's. As was customary

← Solar eclipse viewed in near-totality from the minisatellite Proba-2 on Friday 20 March 2015.

for all scholarly works at the time, *Urania Propitia* was published in Latin, but a German edition was also printed. This meant the book was now accessible to "regular" people, meaning those outside of scholarly and elite circles who did not read Latin. Such publications helped to establish German as an official scientific language.

Historian N. M. Swerdlow wrote that *Urania Propitia* was of "the highest technical level of its age, for its purpose was to provide solutions to difficulties in the most advanced science of the age, the mathematical astronomy of Kepler's Rudolphine Tables."[49] For this singular work, French astronomer Jean-Baptiste-Joseph Delambre called Cunitz a "second Hypatia" and her mythic status even caused her to become known as the Athena of Silesia.

SCIENCE AS THE FAMILY BUSINESS

Cunitz was soon followed by even more women entering astronomy and mathematics in the eighteenth century. While women were still largely prohibited from attending university and denied membership in scientific institutions, those of high social and economic class were offered more opportunities to engage in scientific activities than they were in the Middle Ages. The Enlightenment culture of the eighteenth century encouraged improvement through education including science and mathematics.[50] The latter in particular was seen as being suited to women because it did not require a laboratory full of equipment or a fully-stocked library, both of which required money and access usually reserved for men.[51] Even though women were not trained to be scientists in their own right, they were expected to learn at least enough basic concepts in science and mathematics to engage in interesting social conversation and to help their male relatives and spouses keep accounts for their businesses. For women whose husbands worked in science or astronomy, skills in mathematics were particularly useful.

Nicole-Reine Étable de la Brière Lepaute of France (January 5, 1723–December 6, 1788) was the wife of a royal clockmaker, Jean-André Lepaute, and it was through her work with him that she found a way into the world of science. Their first collaboration was his 1755 book *Traite d'Horlogerie* (*Treatise on the Clock Industry*). She calculated the oscillations of pendulums and pendulum lengths to correspond with numbers of vibrations. Her calculation skills caught the eye of astronomer Jérôme Lalande, one of Jean-André's other collaborators.

Lalande was so impressed by Lepaute that he recruited her to help him solve a particularly difficult problem for the astronomical community—the predicted return date of Halley's Comet.

Predicting the date of the comet's return was a problem that neither Edmond Halley himself nor Isaac Newton seemed to be able to solve. When Halley first took up the puzzle of the comet's return, he relied on Newton's theory of gravity and a new method of mathematics called calculus to chart the comet's path through the sky. Both Newton and Halley postulated that comets move in parabolic motion around the sun, but knowing the shape and direction of this path was only one part of the equation. They also needed to know how fast the comet traveled on its journey, which they knew would depend greatly on the gravitational attraction of Jupiter and Saturn. The comet and the planets formed a three-body problem with each of the three objects influencing the other two.[52] Halley and Newton were able to account for a two-body problem, at best, but they weren't able to find a solution for three.[53] Solving the three-body problem and predicting the comet's return successfully became an unofficial test for Newton's theory of gravity: if the predicted date aligned with the comet's return, Newton was correct. If not, scientists would have to concede that forces other than gravity held sway over the universe.[54]

In 1757, Lalande, Lepaute, and another collaborator, Alexis-Claude Clairaut, set out to do what some of the most celebrated scientific minds of the day had thus far failed to do. From June to September of that year, the three worked morning to night poring over their calculations in the Palais Luxembourg, calculating the comet's advancement along its orbit by degrees. Clairaut performed calculations for the comet itself while Lalande and Lepaute tackled the three-body problem of Saturn and Jupiter.[55] It was Lepaute who was ultimately able to calculate the amount of attraction Jupiter and Saturn exerted on the comet.[56] They predicted that the comet would reach perihelion (the point in the comet's orbit at which it is closest to the sun) between March 15 and May 15 of 1758, correcting the prediction made by Halley, who estimated perihelion would occur the previous year.[57] Their prediction was off by only two days: the comet actually rounded the sun on March 13.

Lepaute continued to collaborate with Lalande on other projects, and even undertook some of her own. In the 1780s and 1790, Lepaute and Lalande published volumes seven and eight of *Ephémérides* (Ephemerides), an almanac published every ten years that predicted the positions of the sun, moon, and planets. Astronomical ephemeris identify trajectories and positions of astronomical objects over time. For nine years, between 1775–1784, Lepaute also assisted Lalande on another ephemeris, *Ephémérides des mouvements célestes*

←

NICOLE-REINE LEPAUTE

Nicole-Reine Lepaute, French mathematician and astronomer who helped calculate the path and return of Halley's Comet between 1757 and 1758.

→

MARIA GAETANA AGNESI

Maria Gaetana Agnesi, Italian mathematician and philosopher who published *Analytical Institutions for the Use of Italian Youth*, the first calculus textbook written in Italian.

(*Ephemerides of celestial movements*). On her own, Lepaute published in 1763 a map of the April 1, 1764 solar eclipse, which charted the advance of the eclipse in quarter-hours. And she published a series of "Observations" from 1759–1774 in *Connaissance des temps* (*Knowledge of the times*), yet another publication of ephemerides and the national almanac of France.

It is, however, for the prediction of Halley's Comet return that Lepaute is best known. She almost did not get credit for it since Clairaut downplayed her part in the calculations and announced them as his own. Outraged by Clairaut's dishonesty, Lalande cut off communication with him for a year, and ultimately publicly recognized Lepaute for her integral role in their collaborative work.[58] Looking back on all their work together, Lalande wrote in his 1803 *Bibliographie*, that "[Lepaute] performed alone the calculations for the position of the sun, moon, and all the planets," and he praised her as "the only woman in France to have acquired a true understanding of astronomy."[59]

TOWARD A CAREER OF HER OWN

Enlightenment Europe was a contradictory place for women; they were encouraged to embrace a scientific education yet were denied scientific careers of their own. As Lepaute's story shows, it was difficult for women to gain recognition for their work, as they were so often eclipsed by male collaborators in their roles as "invisible assistants."[60] For a woman to be visible and autonomous in her scientific work often required a male collaborator such as Lalande to speak on her behalf. Even Cunitz's *Urania Propitia* was suspected to be the work of her husband in the eyes of the astronomical community. This prompted him to write a preface to later editions crediting the work to Cunitz "lest anyone falsely think the work perhaps not of a woman, pretending to be of a woman, and only thrust upon the world under the name of a woman."[61]

Maria Gaetana Agnesi of Milan (May 16, 1718–January 9, 1799) also had the good fortune of a supportive male relative in her father, Pietro. Agnesi exhibited a piercing intellect from an early age that was nourished by Pietro and his hired tutors. Pietro's efforts, however, were not driven purely by paternal love. He wanted to make his way into the aristocracy by catering to the city elites and bolstering his social prestige.[62] In his palazzo, he hosted a "*conversazione*," a gathering of elites for conversation and entertainment in which Agnesi frequently took center stage. She delighted Pietro's guests by carrying on conversations in fluent French at the age of five and delivering Latin orations in defense of women's education by the age of ten. She even discussed Newtonian physics with the prince of Poland in seamless Latin and Italian as a teenager.

Agnesi was struck with an unknown illness in 1730 and gradually began to withdraw from the public life that her father had forced upon her. In 1738, she published *Propositiones philosophicae* (*Philosophical Propositions*), a collection of essays on natural philosophy and science inspired by the *conversazione*, but she had begun to focus specifically on mathematics, which coalesced with her devotion to Catholicism. For Agnesi, mathematics, which she called "the greatest earthly joy available to mankind," was a special kind of knowledge that could reveal absolute truths—a fitting form of study with which to contemplate God.[63]

As a Catholic and a woman of the Enlightenment, Agnesi believed the Church should both embrace elements of modern science and offer better education for the poor and for women.[64] To that end, she published the two-volume *Instituzioni analitiche ad uso della gioventù italiana* (*Analytical Institutions for the Use of Italian Youth*) in 1748, a unique, first-of-its-kind calculus textbook written in Italian, not Latin, for accessibility outside the scholarly elite. *Analytical Institutions* was a complex yet approachable synthesis of algebra, geometry, and calculus, the most advanced branch of mathematics in Agnesi's time.[65] Agnesi dedicated the book to the Austrian empress Maria Teresa, writing a passionate defense of women's education:

> *"I am fully convinced, that in this age, an age which, from your reign, will be distinguished to latest posterity, every Woman ought to exert herself, and endeavour to promote the glory of her sex…"*[66]

BEYOND EUROPE

In writing scientific texts intended for readers outside scholarly circles, Cunitz, with her German *Urania Propitia*, and Agnesi, with her Italian *Analytical Institutions*, laid the foundations for a tradition of women making science available to lay audiences. This practice would stretch well into the nineteenth century, and it was not confined to Europe.

Wang Zhenyi (1768–1797) of China, who mastered astronomy and mathematics, also wrote for students and a lay audience. Zhenyi was born in 1768 during the Qing Dynasty in Jiangning Prefecture (now Nanjing). Like so many other learned women barred from participating in academies and institutions in both China and the West, Zhenyi began her education at home. Her grandfather taught her astronomy and mathematics while her grandmother taught her poetry. Much of her knowledge in astronomy and mathematics, however, was self-taught, a journey that she detailed in the four-volume book *Beyond the Study of Mathematics*. Her posthumous work *First Collection from the Defeng Kiosk* reflected just how widely read she was, showing she had read works on Chinese astronomy and mathematics as well as Western texts such as Euclid's work on geometry in *Elements*.[67]

Both Chinese and European thought influenced Zhenyi's work in interesting ways. She fell in step with Chinese astronomical trends that had long valued celestial prediction. If accurately done by imperial astronomers, such predictions confirmed the emperor's Mandate of Heaven, but if they were inaccurate, this could signal its loss.[68] In *Dispute of the Procession of Equinoxes*, Zhenyi described the procession of the equinoxes and how to calculate their movement, but bucked tradition by demonstrating solar and lunar eclipses as natural phenomena, rather than events open to an interpretation of imperial power, in *The Explanation of the Lunar Eclipse* and *The Explanation of a Solar Eclipse*. She even developed her own working model of a lunar eclipse by suspending from the ceiling a spherical lamp to represent the sun, a mirror representing the moon, and a round table that was the earth. By moving the three into alignment, she could demonstrate how a lunar eclipse is possible when the moon passes through the earth's shadow. She also supported a sun-centered universe during a time when the Tychonic geo-heliocentric model of the universe was still favored by Qing astronomers.[69]

Zhenyi was heavily influenced by Mei Wending (1633–1721), a prominent Qing mathematician. Much like Cunitz with Kepler's *Rudolphine Tables*, Zhenyi rewrote Wending's *Principles of Calculation* to simplify his multiplication and division calculations. Her book, *The Musts of Calculation*, was more approachable to a wider audience of readers.[70] Other accessible mathematics texts for beginners included *Arithmetic Made Easy*, *Simple Calendar Calculations*, and *The Simple Principles of Calculation*. Though few complete copies of her dozens of works have survived, what does remain shows a voracious scientific mind.

Zhenyi along with Cunitz, Lepaute, and Agnesi are not anomalies in the history of science so much as evidence that women have always been capable of succeeding in astronomy and mathematics. The majority of women during this time period did not have those opportunities, either because they were of a lower social class or did not have an encouraging male relative to help make a path for them in science. What heights women could have reached if the world of science had been more democratic and more open at this time is merely guesswork, but what is known from the lives and work of these women is that when given the opportunities, resources, and support to succeed and when their knowledge is valued, women have always had the power to shape science itself.

Maria Cunitz

(circa 1600/10–August 22/24, 1664)

Born between 1600 and 1610 in Silesia, astronomer Maria Cunitz would
eventually become known as the Second Hypatia and the Athena of
Silesia. As a woman, Cunitz was denied formal education, instead being
educated at home by her father Henrich Cunitz, a skilled physician.
Under instruction first from her father, and later from her husband Elias
von Löwen, Cunitz became proficient in mathematics and astronomy
and fluent in seven languages: Greek, Latin, German, Polish, French,
Italian, and Hebrew.

Cunitz and her husband, who she married in 1629, pursued
astronomy together, and she soon mastered higher mathematics.
She easily understood Kepler's *Astronomia Nova* and was one of the
first people to embrace the three laws of planetary motion laid out
in its pages. While exploring his *Rudolphine Tables*, a star catalogue
with directions for calculating planetary positions, she noticed an
error in Kepler's mathematics and corrected it.

In 1650, at her own expense she published *Urania Propitia*, a corrected
and simplified 250-page version of Kepler's *Rudolphine Tables*. She
published it in both Latin, the official language of learned European
natural philosophers and mathematicians, and in vernacular German,
a choice that helped establish German as a language of science.

When Cunitz died on August 22/24, 1664, *Urania Propitia* was
the only work she had published. But since the book was such a
remarkable contribution to astronomy, her legacy endures. In 1960
the minor planet Mariacunitia and in 1961 Cunitz Crater on Venus were
named in her honor.

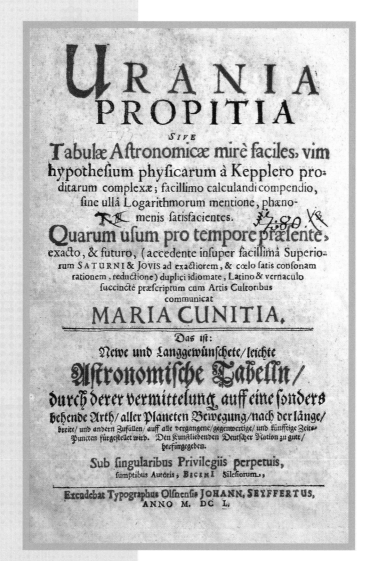

↑ Frontispiece of Maria Cunitz's
Urania Propitia, published in 1650.

CHAPTER 4

The Wives and Sisters of Scientific Partnerships

LOVING ASSISTANT OR CHEMIST?

In a 1788 double portrait of Marie-Anne Paulze (January 20, 1758–February 10, 1836) and Antoine Lavoisier, the French painter Jacques-Louis David immortalized one of the most famous scientific couples in history. The portrait shows the couple at work, with Antoine sitting at a table covered in a lavish scarlet velvet cloth with his chemical instruments before him, pausing from writing his *Traité élémentaire de chimie* (*Elementary Treatise on Chemistry*) to look over his shoulder at Marie-Anne. She stands behind him in a billowing white gown, one hand draped over his shoulder, the other gently resting on the table in front of Antoine, as she stares directly out of the portrait. It appears that she has paused her own work at the drawing tablet behind her to check Antoine's work. Even though Marie-Anne and Antoine each have portraits of their own, David's rendering of the couple in their impressive Parisian home laboratory has become the image most readily associated with the name Lavoisier.

This portrait, titled *Portrait of Antoine-Laurent Lavoisier and his Wife*, has been interpreted in different ways, as both a sign of Marie-Anne's devotion as Antoine's loving assistant, and as evidence of Marie-Anne's equal collaboration in the making of modern chemistry.[71] The most accurate interpretation, perhaps, lies somewhere in between.

Born in 1758 in Montbrison, France, Marie-Anne Pierrette Paulze married Antoine at the age of thirteen in 1771, even though Antoine was fifteen years her senior. Antoine was a wealthy lawyer, but his true vocation was science, particularly chemistry. Antoine was a key figure in the chemical revolution of

← Jacques-Louis David's double portrait of the Lavoisiers, titled *Portrait of Antoine-Laurent Lavoisier and his Wife*, created in 1788.

the eighteenth century, and he is now popularly remembered as the "father of modern chemistry" for overturning the dominant theory of phlogiston by discovering oxygen's role in combustion. It is possible that none of this would have happened without Marie-Anne.

Soon after their marriage, Marie-Anne learned English and Latin to aid Antoine, whose knowledge of foreign languages was weak. Much of the provocative work in chemistry was being done in Great Britain, so Marie-Anne's translations of English chemical texts were essential to Antoine's knowledge about the science of the day and his participation in learned debate. She also asked Antoine to teach her chemistry, and the couple hired tutors to farther her education.[72] With her own chemical knowledge, she picked apart the translations, adding commentary and criticisms of her own as well as prefaces to some published French editions of chemical texts.

According to a contemporary account by Arthur Young, who visited the Lavoisier's home, Marie-Anne worked in the laboratory alongside Antoine.[73] Marie-Anne herself illuminated her role in the laboratory in her own sketches. Two of the surviving drawings depict Antoine and a team of assistants carrying out experiments with oxygen and respiration while Marie-Anne sits at a table diligently observing and recording the experiment. She was no invisible assistant—she very much wanted to be seen. And by inserting herself and representing her role in the laboratory in her artistic works, she ensured that she was.

A skilled artist as well as a chemist, she created thirteen copper plate engravings for Antoine's 1789 foundational *Elementary Treatise on Chemistry*. The meticulously prepared plates, all signed "Paulze Lavoisier Sculptist," captured accurate, to-scale drawings of Antoine's equipment. This realistic rendering of the equipment was a stark departure from other chemical texts of the time, which usually employed an impressionist style for diagrams.[74] Marie-Anne's engravings and the detail they portray were a resource for other chemists who sought to build their own instruments and replicate Antoine's results—and they doubled as significant publicity for Antoine's brand of chemistry.[75]

In 1794, during the French Revolution Reign of Terror, Antoine, who was both an aristocrat and tax collector, was executed by guillotine. Marie-Anne worked tirelessly to restore Antoine's reputation, and she excoriated those who were complicit in the accusations that led to his imprisonment and beheading. Following her own two-month imprisonment in the Bastille, Marie-Anne reclaimed their requisitioned laboratory equipment and other possessions and took it upon herself to publish Antoine's manuscript for *Mémoires de physique et de chimie* (*Physics and Chemistry*

Thesis), which he drafted during his imprisonment. To cement her late husband's reputation as a singular figurehead of modern chemistry, she penned an anonymous preface that attributed to him, and him alone, the new world of chemistry.

By modern standards, it may seem that Marie-Anne played the role of marginal assistant to Antoine's lead, but the Lavoisiers' story represents an egalitarian scientific collaboration between men and women. During the Enlightenment, women and men were seen as different but complementary to each other's opposing fundamental natures: women were the civilizing force that cooled men's aggression.[76] Marie-Anne and Antoine's double-portrait shows the separate but coalescing forms of scientific labor that each undertook, culminating in the work laid out on the table before Antoine, and over which Marie-Anne presides.

FINDING PARTNERSHIP IN THE STARS

The Lavoisiers were not the only scientific marriage, even if they remain among the most famous. Before Marie-Anne was sketching experiments in the laboratory, Elisabetha Catherina Koopman Hevelius (January 17, 1647– December 22, 1693) was running her husband's large private observatory *Stellaeburgum* (Village of the Stars.) Like the Lavoisiers, the collaboration between Elisabetha and Johannes Hevelius of Danzig is also visually immortalized. In an engraving on the frontispiece to Johannes's 1673 *Machina Coelestis*, a huge brass sextant stands in the middle of an open window from which the sky beyond is visible. The sextant required two people working together to take the measurements between stars: operating the instrument on the left is Johannes, and on the right is Elisabetha. It was the first printed image of a woman astronomer at work.[77]

Born in 1647, Elisabetha had always been curious about the stars, and when she was still a child, she sought out Johannes, who had published his famous moon atlas *Selenographia* the same year as her birth. He told her that when she was older, he would teach her about the wonders of astronomy, a promise she did not forget. When his first wife Catherina died, the then fifteen-year old Elisabetha reminded Johannes, aged fifty, of his promise. A few months later, the two were married, marking the beginning a scientific partnership that would last nearly twenty years.

For an intelligent and curious woman, marriage to such a distinguished man of science was an attractive prospect at a time when university education would have likely been out of the question. For Elisabetha, access to one of the most

ELIABETHA HEVELIUS

←

Seventeenth-century engraving
of Johannes and Elisabetha Hevelius
using their brass sextant in their
observatory, *Stellaeburgum*.

→

CAROLINE LUCRETIA
HERSCHEL

Nineteenth-century lithograph of
sibling astronomers William and
Caroline Herschel.

famous observatories in Europe was the next best thing, and she took on the mantle of the scientific wife with a fervor, all while running a household with three children. Within a few years of their marriage, she had taken charge of *Stellaeburgum*, assisting and collaborating with other astronomers who came to work at the observatory and use its first-class instruments. In one account, Elisabetha and Johannes played host to Edmond Halley, then a young emerging astronomer, and Elisabetha even partnered with Halley to instruct him in the use of the sextant.[78]

Like Marie-Anne Lavoisier (see page 47), Elisabetha undertook translations for her husband, translating Latin texts and correspondence with other astronomers. But she also worked on stellar calculations, which were key to the star catalogue and maps that Johannes meant to be an improvement on those of Johannes Kepler. During the seventeenth century, astronomers desired more precise catalogues that accounted for stellar parallax according to the Copernican theory of earth's motion around the sun.[79] But calculating the distance between near and more distant stars was an extremely difficult measurement to quantify, and they were attempting to do so by using only their brass instruments and naked eyes.

After nearly sixteen years of collaborative work, almost all of the Heveliuses' instruments, data, and library were consumed by a fire in 1679, and they were forced to rebuild in order to carry on their work. After Johannes died in 1687, Elisabetha continued the research alone and ultimately published their magnum opus in 1690. The *Prodromus Astronomiae* was a collection of three publications. The first was a catalogue of stellar positions and magnitudes of 1564 stars, 600 newly discovered stars, and about a dozen new constellations. The *Prodromus* also included documentation of the methods they used for calculations, and examples of how Johannes used a sextant and quadrant to calculate a star's longitude and latitude.[80] *Prodromus Astronomiae* was the largest and most accurate stellar catalogue to date, and it was the last to be completed without the aid of the telescope.[81]

Even though the *Prodromus Astronomiae* was without a doubt a joint effort between Elisabetha and Johannes, the book bore only his name as author, as was conventional at the time. In this way, Elisabetha's work was absorbed into that of her husband's, making it difficult to quantify her exact contributions to the published texts. This is a fate that befell numerous scientific wives, many of whom, unlike Elisabetha, remain unknown because they were not given visibility on the frontispieces of their husband's work. Johannes broke with tradition when he included her in *Machina Coelestis*, not only in visual form but also in writing by attributing particular observations to her, "[m]y dearest wife."

Not all scientific partnerships between men and women were marriages; sisters and brothers also worked together. One such little-known sister–brother collaboration of the Scientific Revolution was that of Lady Katherine Ranelagh (March 22, 1615–December 3, 1691) and her brother Robert Boyle, a towering figure in chemistry, philosophy, and modern experimental science. Twelve years his senior, Lady Ranelagh exercised more of a mentorial influence on Boyle rather than the hands-on role of experimentation and computation played by Marie-Anne Paulze Lavoisier and Elisabetha Koopman Hevelius. Without a visual embodiment of their partnership, Lady Ranelagh has languished as a footnote to Boyle's life and career.

Lady Ranelagh was one of the most respected and influential women in seventeenth-century England. Though her personal life was often unhappy, marked by an unfulfilling arranged marriage to and ultimate separation from Arthur Jones (later named Viscount Ranelagh), she was well-connected with prominent intellectuals and parliamentary politicians of the time, particularly as a member of the Hartlib Circle. Initiated by Samuel Hartlib in 1630, the Circle was a correspondence network of politically engaged and scientifically curious European intellectuals, including the writer John Milton, William Laud the Archbishop of Canterbury, and eventually Boyle himself. As a teen, Boyle corresponded frequently with his older sister, and it is through these letters that Lady Ranelagh's influence on a young Boyle's intellectual and moral development is apparent.[82]

While forming his philosophical ideology, Boyle sought Lady Ranelagh's guidance and he often utilized her own anecdotes in articulating his ideas, sometimes referring to her as "a Sister of mine." In one of his published philosophical texts, *Occasional Reflections Upon Several Subjects*, Boyle dedicated the entirety of the work to his sister, calling her "Sophronia," a Greek word for temperance and wisdom. Boyle, however, never attributed Lady Ranelagh by name in his published writing to preserve her modesty, which has made it difficult to pinpoint her exact influence on his work.[83]

The two shared a growing enthusiasm for chemistry. Lady Ranelagh read Boyle's drafts of chemical texts and encouraged him to publish his works, and they shared chemical recipes and experiments in their correspondence. Lady Ranelagh was skilled in chemistry as it applied to medicine, which she saw as the branch of natural philosophy with the most practical benefit to society.[84] She was so respected for her medical recipes that Boyle often cited her as an authoritative source in his own work, particularly in *Usefulness of*

Natural Philosophy, referring to her as "a great Lady."[85] When Boyle moved in with his sister, where he lived for the next twenty three years until their nearly simultaneous deaths in 1691, Lady Ranelagh set up his home laboratory with everything he needed.

In 1660, the Hartlib Circle in which Lady Ranelagh enjoyed significant influence dissolved and the Royal Society was officially formed, absorbing many members of the Hartlib Circle along with Robert Boyle as a founding member. As a woman, Lady Ranelagh was not considered for membership in the Royal Society, despite being deeply immersed in the same intellectual milieu and carrying on collaborative relationships with Royal Society members.[86]

Because of the various factors that kept Lady Ranelagh from visibility— Boyle's pseudonyms for her in his works and the exclusion of women from the Royal Society—her part in the Scientific Revolution can be easy to miss. Her intellectual force, however, is embedded into Boyle's own work in ways that may never be quantified. What is more, the ideas that she helped to cultivate through her relationships with pivotal members of the Hartlib Circle and later the Royal Society formed part of the Society's foundation, and it still stands today.

EVENTUAL INDEPENDENCE

In contrast to Lady Ranelagh, Caroline Lucretia Herschel (March 16, 1750– January 9, 1848) has received a fair amount of recognition as one half of the most well-known scientific sibling partnership with her older brother William. Despite a long list of accomplishments and astronomical discoveries, Caroline had little agency and independence over her own life, a circumstance she recognized herself when she said, "I am nothing. I have done nothing; all I am, all I know I owe to my brother. I am only the tool which he shaped to his use."[87]

As with many women in the eighteenth century, Caroline had limited education, an obstacle to success that was compounded by a childhood filled with active neglect that bordered on abuse.[88] William, closest of her two brothers, removed her from the family household in Hanover, Germany, and employed her as his housekeeper and assistant in England, something for which she always felt indebted to him. Initially, William wanted to pursue music, with Caroline working with him as a music copyist and a singer. When William's interests changed to astronomy, he simply enlisted Caroline as his assistant astronomer without consulting her. In her memoir, she recalled, "and by way of encouragement a telescope adapted for 'sweeping'…was given me."[89]

In her new assistant role, Caroline recorded William's observations, polished his telescope mirrors, and copied astronomical catalogues. She even fed and read to William while he worked, tasks far outside the realm of an assistant by any standard. In 1786, William published the *Catalogue of Nebulae and Clusters of Stars* in the *Philosophical Transactions*, and despite Caroline's diligent assistance in its composition, the published catalogue bore only William's name, reifying her place as an assistant. After William died, however, Caroline reconfigured the catalogue's nebula and stellar clusters into zones, a colossal undertaking that required the reconfiguration of 2,500 nebulae. The revised catalogue was renamed the *New General Catalogue*. For this, the Royal Astronomical Society (RAS) awarded her their Gold Medal in 1828.

Through her own observations, Caroline discovered eight new comets between 1786 and 1797. Five of these were published in the *Philosophical Transactions* of the Royal Society, making her the first woman to publish there. In addition to her comets, she also discovered three new nebulae and identified 560 stars that had been overlooked by John Flamsteed, the first Astronomer Royal, in his authoritative catalogue of stars. In 1798, she published the *Catalogue of Stars, taken from Mr Flamsteed's Observations*, which contained the additional

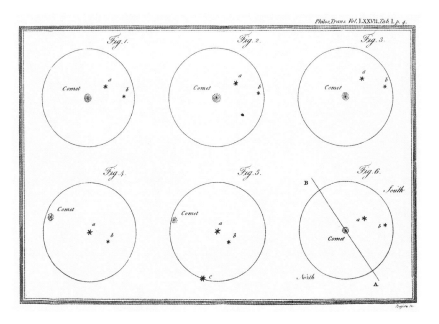

Philos.Trans. Vol. LXXVII. Tab. I. p. 4.

↑ Diagrams show six different positions of a comet discovered by Caroline Herschel in August of 1786, published in *Philosophical Transactions* on 1 January 1787.

stars and an index she created to allow users to easily check observations against their entry.

Caroline received honors both from the RAS as an honorary member and from the King of Prussia, who awarded her a Gold Medal for Science. Despite the laudatory honors, Caroline did not take much pleasure in them, writing in her memoir upon receiving her honorary membership from the RAS, "God knows what for." She seems to have taken the most pleasure in earning a stipend of £50 per year from the crown for assisting William, making her the first professional woman scientist in Britain. It was one of the few times in her life she was able to exercise some form of independence: "...the first money I ever in all my lifetime thought myself to be at liberty to spend to my own liking."[90]

RELEGATED BY HISTORY

The romanticized story of the scientific marriage in which a loving wife selflessly assisted her famous husband has been uncritically handed down for centuries. The women in these partnerships, no matter how skilled or intelligent, have been relegated to the role of a helpful assistant, a position that epitomized a woman's secondary place to men in society at large.[91] But the knowledge these women had and the work they did was integral to the published works of their husbands and brothers and to the development of modern science and its institutions. It is impossible to disentangle the contributions of these partnerships from one another in an attempt to quantify and rank who did more because it was work undertaken equally, collaboratively, and toward the same end.

ANDREAE VESALII
BRVXELLENSIS, SCHOLAE
medicorum Patauinæ professoris, de
Humani corporis fabrica
Libri septem.

CVM CAESAREAE
Maiest. Galliarum Regis, ac Senatus Veneti gra-
tia & priuilegio, ut in diplomatis eorundem continetur.

BASILEAE.

CHAPTER 5

Women and the Science of the Body
in the Scientific Revolution

THE OTHER SCIENTIFIC REVOLUTION

Historians have described the period from the mid-sixteenth century to the mid-eighteenth century as the most important and productive time in the history of science. The two centuries encompassed many important scientific advances, such as Nicolaus Copernicus's explanation of the heliocentric cosmos in 1543 and William Harvey's 1628 publication of his theory of the circulation of blood. Because of its many field-defining discoveries and breakthroughs, this early modern period is often referred to as "The Scientific Revolution." The popular understanding of the Scientific Revolution is one that encompasses the lives and work of famous and prestigious men. Women's opportunities to participate in the emerging culture of modern science were limited, excluded as they were from the laboratories and universities that were emerging in Europe and around the world. In more recent decades, however, feminist historians have worked to recover the role of women in this rich period of scientific discovery.

One of the most important ways historians have recovered the relationship of women to science in this period is *visually*. During the Scientific Revolution, the culture of printed books allowed for faster transmission of knowledge and, crucially, more easily reproducible, complex illustrations. The intensely visual culture of the early modern period is one of its hallmarks, and nowhere is this visuality more important—both as a medium for transferring and *creating* knowledge—than in the science of anatomy. Through the creation and

← The frontispiece from *De humani corporis fabrica* (*The Fabric of the Human Body*) by the Italian anatomist Andreas Vesalius, published in 1543. The woodcut scene depicts a public anatomy, in which Vesalius dissects the corpse of a woman criminal.

transmission of both two- and three-dimensional images of the human body, early modern anatomists created a densely visual understanding of the human body, one with which the social and cultural place of women was intricately intertwined. Women sought knowledge of their own bodies and, in turn, used this knowledge to heal other women. As doctors and anatomists, women who navigated the prohibitions against their study of nature alter our understanding of the male-dominated narrative of the Scientific Revolution.

THE EARLY MODERN IMAGE OF THE BODY

In 1543, the same year that Copernicus published his controversial *De revolutionibus orbium coelestium* (*On the Revolutions of the Heavenly Spheres*), an anatomist and, at times inveterate grave robber, named Andreas Vesalius published the most magnificently illustrated book on anatomy in the Western world, *De humani corporis fabrica* (*The Fabric of the Human Body*). The original folio edition was 17 inches (43 cm) tall, over 700 pages long, and contained detailed woodcut illustrations of the dissected human body. Enormously expensive to print, the book would have been owned only by the wealthy and learned. The frontispiece depicts a raucous public anatomy demonstration, in which Vesalius himself dissects a body in front of a huge crowd of spectators. On the dissection table is the only woman in the scene, an unnamed prisoner whose uterus is exposed to the gawking party.

Although the familiar image of the study of anatomy is perhaps like that of the frontispiece of Vesalius's book, filled with eager students and a distinguished professor demonstrating on a corpse, the practice of anatomy is not rooted in learned university medicine but rather in public health, religious practice, and the knowledge of women. Historian Katharine Park has documented the religious practice of autopsy and preservation of relics from dissected holy women and the dissection of wealthy women who died in childbirth and were autopsied to find the cause of their illness. Both of these dissection practices took place outside of academic anatomy, but both were deeply concerned with the mysterious nature of the female body and the value of the secrets that might be held within. Women seemed to possess hidden knowledge about sexuality and generation and the unseen inner structures of the body that remained unknown to both men and women.[92]

Anatomy is one of the scientific disciplines in which the complex interplay between women as practitioners and as subjects of natural inquiry is most apparent. Women have always been *knowers* in anatomy, in the sense that just like everyone else, they have bodies about which they possess intimate knowledge. And women have historically created and used knowledge about the bodies of other women with whom they had kinship, community, or professional relationships. The female body was of particular interest to early modern anatomists, nearly all of whom were men. The secrets of reproduction, the intimate knowledge shared between women, and the mysterious interior structures of the body regulating both, were among the critical objects of inquiry in this science.

Women who pursued knowledge of nature were aided in many cases by influential members of their societies who bent or changed longstanding rules about women's participation. In Italy, women were central to the revived intellectual life of its cities during the Enlightenment. Born in Bologna, in 1714, anatomist Anna Morandi lived during a period of intense change for the city. Government and church officials had begun extensive reform aimed at recapturing Bologna's former cultural and intellectual glory. In the Middle Ages, Bologna had been the home of some of the most important figures in the science of anatomy, including Mondino de Luzzi, said to be the first person in Europe to demonstrate anatomy using a real corpse.[93]

ANNA MORANDI'S WAX BODIES

Anna Morandi (January 21, 1714–July 9, 1774) married into the anatomical profession when she wed the artist Giocanni Manzolini in 1740. That same year Manzolini began an apprenticeship with the celebrated anatomist and wax sculptor Ercole Lelli. Manzolini came from a long tradition of wax sculpting, which was often used to create religious images but which also became important for teaching anatomy and dissection in lieu of human corpses. In 1745, Morandi began to assist her husband and teach anatomy classes to medical students in their home studio; their partnership and the excellent quality of their work earned them international renown.[94]

In 1749, a professor at the University of Bologna commissioned the couple to create models for an anatomy school he was opening in his home, which was later moved to the Institute of Science. Morandi and Manzolini created some twenty models of the pregnant uterus and the female reproductive system that were used to teach obstetrics.[95] When Manzolini died suddenly in 1755, Morandi took over his anatomical modeling practice and in his absence, her career flourished.[96]

According to scholar Rebecca Messbarger, Morandi's status as a widow made her something of a curiosity in Bologna, given the nature of her work.

← **ANNA MORANDI MANZOLINI**

Anatomist Anna Morandi Manziolini studied anatomy and wax modeling with her husband, and took over his practice after her death. She became very famous in her time, and many of her exquisite wax anatomical models survive today.

→ **DOROTHEA ERXLEBEN**

The first woman to attain a medical degree in Germany, Dorothea Erxleben advocated for women's education and fought to use the title "doctor" in her medical practice.

← **LAURA BASSI**

The second woman to earn a doctoral degree, Italian physicist Laura Bassi taught at the University of Bologna and popularized Newtonian mechanics in her home country.

But the city's intellectual leaders also feared she might leave, taking her unmatched talents with her.[97] To secure her services to the intellectual life of the city and the University of Bologna, the pope awarded her a lifelong stipend. In 1758, she was inducted into the prestigious Clementine Academy of the Arts, and two years later she accepted an appointment as the Chair of Anatomical Modeling at the University of Bologna.[98] Messbarger points out that Morandi's work, while virtuosic, was not merely about making art or even making teaching models for other people. Morandi was herself a very skilled anatomist, who developed new dissection techniques and conducted influential observational and theoretical work in the course of making her wax figures. She compiled her notes on the body into a 250-page book that explained both the form and the function of the parts of the body.[99]

Scholars have noted the innovative features of Morandi's work and how her models departed from certain traditional ways of representing the human body in three dimensions. Rather than creating a merely rational medical image of the inert body, Morandi's models of sensory organs and hands aimed to capture the physiological processes of the body as well.[100] In her notes about the anatomy and physiology of the hand, Morandi wrote not only about its parts and their appearance and organization, but of the process of touch itself:

> "The sense of touch permeates the entire human body, each part being possessed of it and furnished with many nerve fibers that renders it sensitive to any contact with a foreign body and in accordance with the delicacy or harshness of the body that surprises it, it readjusts and moves."[101]

Morandi's understanding of the human body made her much more than an artist or craftsperson. And her wax models are more than just representations intended for the instruction of acolytes. They are the physical expression of her research, not at all unlike Vesalius's now-mythic *De humani corporis fabrica*.

Morandi died in 1774, and her legacy has since passed into relative obscurity. Some of her models are conserved in the Museum of Normal Human Anatomy at the University of Bologna, but others are in storage and in poor condition.[102] Messbarger argues in particular that her early biographers did her a disservice in implying that her talents were simply innate or instinctual, rather than the product of extensive research and dedicated study.[103] In reality, Morandi was an accomplished scientist, famous in her own time, whose story complicates popular ideas about women's role as creators of anatomical knowledge and the scientific establishment in Enlightenment Europe.

The eighteenth century was a moment of transition for women in science.[104] Having long been participants in home workshops, some women in the early modern period were able to secure places at universities, participating in the creation of knowledge about the human body as doctors and anatomists. In the eighteenth century, Dorothea Erxleben (November 13, 1715–June 13, 1762) earned a medical degree at Halle, a university in Germany, and was the first woman in the country to do so.[105] Much like women who received training at home as part of family businesses, many wealthier women received informal scientific training, but to be recognized for this work, they needed certification from a public institution like a university.[106]

Initially educated by her family, Erxleben received lessons similar to those of her brother at her father's insistence.[107] When Erxleben was a teenager, she began working with her father Christian Polycarp Leporin, a physician at Quedlinburg in Prussia, and she later petitioned the King of Prussia to let her study medicine and receive certification at university.[108] She wrote a treatise on the institutional and cultural barriers women faced in receiving education, which was published in 1742, around the same time that she finally received permission to enroll at Halle. Her studies were interrupted by the Prussian-Silesian war, during which she ran her father's medical practice in his absence. After his death, she took over the practice while also raising her children and was unable to return to her doctoral studies until she was thirty-seven.[109]

The cultural context in which women pursued scientific careers matters a great deal. In Erxleben's case, she had to actively persuade the Prussian government to let her study, but she was among a number of noblewomen who had been granted permission.[110] While there were no universities for women in Germany at the time, there had been proposals for such, and Erxleben herself knew of learned women in other countries.[111] She still faced opposition from men who maintained that it was inappropriate for women to be educated at all, let alone in the same schools as men.

In 1753, three doctors in the city of Quedlinburg lodged a formal complaint about "quackery" ruining the profession, citing Erxleben as the principal offender. These complaints resulted in a new city ordinance forbidding anyone but a licensed doctor to dispense medical care. Having not yet finished her degree, Erxleben was prohibited from practicing. In a written response to the charges of "quackery," she distinguished herself from real "quacks" who, unlike herself, had no medical training at all. Later that year, she requested permission to take the final examinations for her degree. The university rector who reviewed her case decided to admit Erxleben to Halle, a decision that signalled early change in German society regarding the education of women.[112] After having practiced medicine publicly for nearly a decade, Erxleben could in 1754 finally use the title *doctor*.

REVOLUTIONS IN MODERN SCIENCE

Both Erxleben and Anna Morandi lived in a time of transition for women in science, alongside contemporaries such as Laura Bassi (1711–1778), a Bolognese philosopher who was the first woman to hold an academic appointment in physics at any university. If anatomy in the late Middle Ages was marked by intensive study of the female body by male scientists, these eighteenth-century women prove that this field was not solely the domain of men. Many women would follow in Erxleben's footsteps to become doctors and healers, often facing the same kind of social and legal challenges. Morandi's path was even more unusual. Being a doctor was one thing, as medical care of family and neighbors had long been the purview of women. But being an anatomist and cutting open dead bodies to peer inside them, not to heal but to understand their complex workings, had been a privilege reserved for men since ancient times.

Morandi is at least as important to the history of the so-called "Scientific Revolution" as Vesalius, if not more. Her work was revolutionary in scientific terms, but in social terms as well. Vesalius's prestige and fame were in part due to his magnificent book but also to his social station, access to education, and funding for his work—all of which were available to men who showed themselves to be talented enough. The same cannot be said of most women in this period, and thus, Morandi's success as an anatomist made her an outlier. She leveraged her social position and her connections to influential scientists and reformers in Bologna to build an extraordinary career as an anatomist.

The medical women of the eighteenth century pursued their craft in the manner available to them as women, but their accomplishments are no less profound than those of their male counterparts. In the early modern world, it seemed that the deepest secrets of the cosmos were being laid bare, from the organization of the heavens themselves to the intricate workings of the smallest structures of the body. The revolutions of this part of the history of science were not merely the discoveries of individual visionary men but also the nascent changes to the access women had to education and professional opportunities. People like Morandi and Erxleben were extraordinary, but their success was also a sign of the changing nature of science itself.

CHAPTER 6

Empire and Exploitation in
the Age of Exploration

COLLECTING THE WORLD'S NATURAL WONDERS

In 1766, two ships, the *Boudeuse* and *Etoile*, set sail from Nantes, France under
the command of Louis Antoine de Bougainville, charged by King Louis XV to
explore the Pacific and take possession of any new or empty land in the name of
the French Empire. While Bougainville settled at the helm aboard the *Boudeuse*,
a short, smooth-faced valet named Jean Baret (July 27, 1740–August 5, 1807)
was setting up his lodgings on the *Etoile* in the cabin of his employer, Philibert
Commerson. This was an unusual arrangement, but Baret and Commerson
reasoned to the rest of the crew that since they both suffered from severe
seasickness they should share a cabin at night. This arrangement was not nearly
as unusual as the real story: Jean Baret was actually Jeanne Baret, not a man but
a woman disguised as a man to gain passage on the expedition.

Commerson was the expedition's naturalist, employed personally by the
King, and as such, the crown gave him 2,000 livres to pay an assistant and
illustrator.[113] Commerson's assistant, however, was a woman, and the French
naval ordinance of 1689 forbade women from spending extended periods
of time aboard naval vessels—staying overnight or for the duration of the
expedition was completely out of the question. To continue working together,
Baret's disguise was the only option.

The exact nature of Baret and Commerson's relationship is somewhat
shrouded in mystery. Prior to the 1766 expedition, Baret worked as
Commerson's paid assistant in Paris for two years, and before that, Baret

← Maria Sibylla Merian's rendering of *Flos pavonis*, the peacock flower, and the life-cycle of a pale
sea-green caterpillar in her 1730 book *Metamorphosis insectorum Surinamensium*.

moved into Commerson's house as housekeeper shortly after the death of his wife. Baret was five months pregnant, which has raised questions about the child's parentage, especially since Baret refused to name parentage in legal documents. One historian speculates that the two met and began a romantic relationship in the Loire Valley where Commerson would roam about botanizing and where Baret would collect plants and herbs to supply medical men and apothecaries.[114] Whatever the nature of their partnership, it is clear that they did not want to be separated.

As part of the expedition, Baret traversed the Pacific, visiting places that few European women had set foot in, especially those born into the laboring class like Baret. Voyaging naturalists of the upper classes often hired lower-class assistants for overseas exploration, but they did not often employ women in this capacity.[115] As Commerson's assistant, Baret carried supplies during their field excursions, collected specimens of flora and fauna, and amassed herbaria of foreign plants from across the world from Rio de Janeiro to Madagascar. In the eyes of others, Baret was not just an assistant—Bougainville referred to her as an "expert botanist."[116] The National Museum of Natural History in Paris records 1,735 specimens collected by Commerson.[117] And even though the specimens are listed under his name as the primary discoverer, Baret as his close assistant would have been part of their collection and preservation.

The three-year voyage became the first French circumnavigation of the globe. As part of the crew, Baret, in the disguise of a man, became the first woman to accomplish the feat. Even though the accounts of those onboard differ as to when Baret's identity was officially discovered, it is clear that the crew did become aware of her deception during the expedition. The crew returned to France in 1769, but Baret stayed with Commerson, only parting from him when he died on Isle de France, now named Mauritius, in the Indian Ocean in 1773. When Baret returned to France some years later, she did not escape scrutiny for her actions on the expedition. She had broken naval code, but Bougainville vouched for her. Instead of dispensing punishment, the royal navy paid her a pension of 200 livres per year starting in 1785 and referred to her with an honorific title—"femme extraordinnaire."[118]

Baret's story is indeed extraordinary, as it demonstrates the extreme lengths women had to go to in order to break into fields that were exclusively the domain of men. Baret's story, however, cannot be separated from the darker story of French imperialism and colonization. She participated in an expedition to claim lands for the French empire, and whether those lands were the homes of Indigenous people did not matter to France. After the loss of the

↑

JEANNE BARET

Disguised as a man to gain passage on a French vessel, Jeanne Baret
became the first woman to circumnavigate the globe.

colonies in its defeat in the Seven Years War, France was attempting to reclaim its diminished power on the world's stage. Commerson's and Baret's roles as naturalists on a military vessel were an important part of the Navy's colonial mission, for the expansion of Europe's military might went hand-in-hand with its scientific exploration. The collection of the natural wonders of foreign lands and their eventual integration into European taxonomic systems, such as Carl Linnaeus's system of classification, was a symbol of Europe's dominion over the rest of the world. The commercial trade of exotic plants for food, medicines, and dyes created economic power for European countries. Particularly in this era of diminished French sovereignty, botany was thought to hold the key to stimulating the economy and establishing the empire as a premier European power.[ ]

USING INDIGENOUS KNOWLEDGE FOR EUROPEAN SCIENCE

Commerson and Baret were not the first to take advantage of imperial power to advance their science. In the eighteenth century, ever-expanding European trade and military routes already stretched across the globe. Nearly seventy years before Baret donned her disguise and set sail, Maria Sibylla Merian (April 2, 1647–January 13, 1717) had traveled to Latin America to observe insects in their natural environment. Merian, though, did not disguise herself; instead, to work around the various barriers that kept women at home, she financed her own voyage across the Atlantic in pursuit of her science by selling illustrations and subscriptions to the book she planned to publish after her expedition. She remains the only woman in her time known to have done so.

From childhood, Merian showed a curiosity about plants and insects, but it was her skill as an artist that ultimately lent her a foothold in science. Born into a family of German artists in Frankfurt am Main in 1647, she began her training under her father Matthäus Merian, and after his death, continued under her stepfather Jacob Marrel. Recovered journals of her teen years indicate Merian's illustration style had nearly fully developed by the age of thirteen, and that her affinity for the natural sciences budded in parallel with her art. Beginning in 1660, she documented her experiments on breeding caterpillars and their diets. Alongside her observations, she included a watercolor illustration depicting the caterpillars' life, from egg to excrement. In observing insects directly from real life, Merian was different from other naturalists, who typically studied dead and preserved specimens. Her diligent observation was very fruitful. She documented the metamorphosis of silk worms in her journal nearly a decade before Italian biologist Marcello Malpighi published his account, which has been widely accepted as the first, in 1669.<?>

Merian nurtured her artistic skill and further developed her keen eye for scientific precision throughout her life, even as she became a wife to Johann Andreas Graff in 1665 and a mother to daughters Johanna Helena and Dorothea. In 1675, Merian published the first installment of *Neues Blumenbuch*, a three-volume series, each containing twelve plates of illustrated flowers and insects. In 1679, she returned to her fascination with caterpillars in *Der Raupen wunderbare Verwandlung und sonderbare Blumennahrung* (*The Caterpillar's Marvelous Transformation and Strange Floral Food*), published in two volumes. The book

The collection of the natural wonders of foreign lands and their eventual integration into European taxonomic systems, such as Carl Linnaeus's system of classification, was a symbol of Europe's dominion over the rest of the world.

continued in the style she had begun in her journals many years before: a description of the insects' lifecycle and habitat brought to life with beautiful, precise illustrations. In 1691, Merian divorced Graff, a rare move for a woman at the time, and moved with her daughters to Amsterdam.

Amsterdam was the fulcrum of global commerce for the Dutch and held collections of natural curiosities and wonders from around the Dutch Empire, which encompassed large swaths of the East and West Indies. Drawn to these collections, Merian sought out the city's mayor, who as the director of the East India Company had amassed a large natural history collection. But accustomed as she was to observing live specimens, Merian was disappointed in the collection. She decided to go to the source: "This all resolved me to undertake a great and

↑

MARIA SIBYLLA MERIAN

German naturalist and botanical artist, Maria Sibylla Merian made
significant contributions to our understanding of ecology.

expensive trip to Surinam [sic] (a hot and humid land) where these gentlemen had obtained these insects, so that I could continue my observations."[121]

In the Netherlands, scientific expeditions often followed along Dutch trade routes, but women were not hired by trading companies or funded by scientific institutions. Many in Europe believed that women faced more dangers than men on voyages to foreign lands, and many of their fears revolved around women's reproduction. Some physicians claimed that traveling south of the equator wrought infertility, while others feared that white women in warm climates would suffer heavy menstruation, inducing fatal hemorrhages in the uterus.[122] But Merian was determined, and having earned sufficient funding, in 1699 at the age of fifty-two, Merian and her daughter Dorothea set sail for the Dutch colony of Suriname.

The Dutch colonized Suriname in 1667, and established an economy based on coffee, cotton, sugarcane, and cocoa sustained by the enslavement of Africans and the native Arawak people. Suriname was a notoriously brutal slaveholding regime, and these were the conditions in which Merian carried out her scientific studies. Merian spent twenty-one months in Suriname, observing a variety of live insects, many from the interior of the colony that had yet to be classified, and illustrating their lifecycles in and around the plants on which they feed. The resulting book *Metamorphosis insectorum Surinamensium* (*Metamorphosis of the Insects of Surinam*) was published in 1705, after her return to The Netherlands. Depicting the lifecycles of the insect and the plants among which they lived and fed, Merian's book was an unprecedented study of an interconnected ecology of the natural world. With its sixty beautifully engraved plates and study of newly classified insects, *Metamorphosis* has become a famous and beloved masterpiece of natural history.

Merian uncovered so many new specimens largely because she enlisted the labor of local enslaved people who guided her into the Suriname interior and shared their knowledge about the specimens. In one of her illustrations, Merian rendered a pale sea-green caterpillar living on the peacock flower *Flos pavonis*, a striking flower of vibrant reds and yellows. The peacock flower, Merian was told by enslaved women, was an abortifacient that they used to "facilitate abortions [...] they were unwilling to have their own children become slaves only to have them share in the same miserable existence as they, living under equally terrible conditions."[123] Merian's depiction of the native plant is both a first-hand account of the colony's brutality to enslaved people and the ease with which European scientists like Merian used and exploited the knowledge of the people they enslaved. She even took an enslaved woman home to The Netherlands with her who remains unnamed, referred to only as "my Indian" in Merian's writings.[124]

The international slave trade that made Merian's work possible would eventually come to an end across Europe in the nineteenth century, but its vestiges remained. Classifying nature's flora and fauna under one dominating system remained a priority for empires and nations. Charting unknown, and sometimes dangerous, foreign landscapes infused exploration with a particularly masculine flare that symbolized a passage into manhood for European men.[125] In the nineteenth century, however, more European women traveled abroad than in Merian's time. For British women in particular this was the result of Britain's craving for natural history, much of which was written and published by women, and the still expanding British empire. When naturalist and writer Sarah Bowdich Lee (September 10, 1791–September 22, 1856) arrived on Africa's Gold Coast in 1816, over a century after Merian sailed to Suriname, science was still an imperial endeavor.

A pregnant Bowdich Lee first traveled to Cape Coast Castle in modern Ghana to be with her husband, Thomas Edward Bowdich. Thomas was a junior officer in the Africa Company and a writer with the Company of Merchants who was tasked with negotiating a peace treaty between the British at Cape Coast Castle and the native Asante kingdom. Cape Coast Castle was a vestige of the international slave trade, which had been abolished in the British Empire in 1807. It was one of dozens of "slave castles" that dotted the African coast, having housed in its dungeons captured Africans awaiting the Middle Passage and enslavement abroad. The African Gold Coast held the largest cluster of European fortresses in the non-European world.[126] It was here, on the literal ruins of slavery, that Bowdich Lee made her first contributions to science.

While on the Gold Coast, Bowdich Lee undertook a variety of natural history studies of the local flora and fauna, making her the first European woman to systematically collect plants in tropical West Africa. When Bowdich Lee and and her husband returned to England, he published *Mission from Cape Coast Castle to Ashantee* in which he included the negotiations with the Asante people, an anthropological account of the kingdom, and geographical and natural history observations. Many of the natural history studies were his wife's, but they were incorporated into the published book that bore only his name.

Next, the couple planned a scientific exploration of Sierra Leone, which was founded as a settlement of free Africans after the British abolished the slave trade. To launch their independent exploration, the couple moved to Paris in 1819 so Thomas could train in the natural sciences and the couple could fund their voyage. They met and studied with prominent naturalist and zoologist

Georges Cuvier and earned money by translating natural history texts into English. Bowdich Lee herself translated several works of Cuvier's, including *Elements of Conchology* and *The Ornithology of Cuvier*. During this time, Bowdich Lee published her own book, *Taxidermy: Or, The Art of Collecting, Preparing and Mounting Objects of Natural History*. The book went through six editions, all published anonymously, save for the last.[127]

When they had enough money, the couple set sail for Sierra Leone in 1822, but they stopped in Madeira because Bowdich Lee was pregnant. After giving birth to their daughter, Eugenia Keir, the family moved on to Bathurst on the Gambia River. However, they never made it as far as Sierra Leone because Thomas died in 1824. She stayed in Africa for another two months, and on the return journey, all of her specimens were destroyed. Even Thomas's death and the loss of her materials did not stop her from working, and she published Thomas's last manuscript, *Excursions in Madeira and Porto Santo* in 1825. Much of the work, however, was her own, including an index that documented twenty new species of fish that she had identified. In the French translation of the work, Cuvier added classification notes, and Alexander von Humboldt contributed an appendix applauding the book's observations.[128]

Despite Bowdich Lee's achievements, she was in poor financial straits without her husband's income, so she continued to work writing popular natural history articles and books. Among her books, and perhaps her greatest achievement, was *The Fresh-water Fishes of Great Britain*, published between 1828 and 1838 in twelve installments. In addition to being a beautiful work of ichthyological watercolors, *The Fresh-water Fishes* was a cutting-edge work in classification of fish, incorporating the latest developments in the field by Cuvier, which had yet to be published in Britain. Although she began her career in science in partnership with her husband, she thrived on her own, publishing twenty books over the course of her life.

Unlike Baret and Merian, Sarah Bowdich Lee was openly against slavery. One of her fictional works, "The Booroom Slave," about the horrors of slavery, was even used by abolitionists to expose the inhumanity of the institution.[129] Yet, it was still the vestiges of the slave trade, enabled by imperial expansion, that facilitated her scientific exploration, just the same as the women who came before her. All three of these explorers evaded the rules that kept women on the sidelines of science and discovery, but each of their stories is intricately tied to histories of brutality and colonial extraction.

Section III

The Long Nineteenth Century

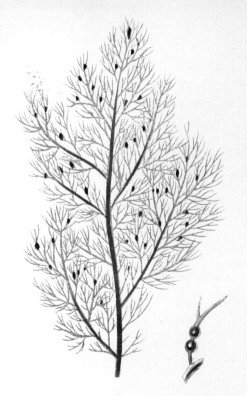

195.—Cystoclonium purpurascens, *Harv.*

196.—Gracilaria multipartita, *J. Ag.*

197.—Gigartina pistillata, *Lamour.*

198.—Gracilaria confervoides, *Grev.*

CHAPTER 7

Women Science Writers and Popularizers

When writer Florence Fenwick Miller (November 5, 1854–April 24, 1935) weighed the state of women in Britain in 1884, she saw progress. Even though they had a long way to go to achieve liberation, more women participated in education, political debate, and social reform than had ever before while many obtained careers and income that had previously been out of their reach. When Miller looked to the future, she saw more progress still. "Mind will be more and more valued and cultivated, and will grow more and more influential; and the condition of women must alter accordingly," she wrote. "Some people do not like this fact; and no one can safely attempt to forsee all its consequences; but we can no more prevent it than we can return to hornbooks..." This new-found value in women's minds, according to Miller, went hand-in-hand with technological progress, particularly technologies that served the written word. Listing them, she wrote, "The printing-press, which multiplies the words of the thinker; the steam-engine, which both feeds the press and rushes off with its product, and the electric telegraph, which carries thought around the globe, make this an age in which mental force assumes an importance which it never had before in the history of making."[130]

The nineteenth century did, indeed, see an unprecedented proliferation of women writers that coincided with the introduction of new printing technologies. All aspects of the printing and publishing process gradually became faster and more efficient. The steam press, which England implemented in 1814, could produce 1,000 to 2,000 pages per hour, whereas its hand-press predecessor was limited to 200 to 400. In 1884, the same year Miller was writing, the rotary-press,

← Plate XLIII from Volume II of Margaret Gatty's *British Sea-Weeds*, published in London in 1872.

powered by electricity, was introduced, generating nearly 20,000 copies per hour. At the same time, literacy rates rose, and the demand for literature of all types—books, newspapers, periodicals—increased in tandem with production. With the high output of printed materials and the increase in demand among readers, publishers turned to women writers out of necessity, and women seized the new opportunities open to them.

Changes in publishing and the literary marketplace had profound effects on the production and distribution of information about science and natural history. In 1815, there were about ten commercial magazines specializing in natural history, science, medicine, and technology, but by 1830 the number had tripled.[131] Between the 1840s and 1850s, around the time the book publishers began utilizing the steam press, the number of science books quadrupled from that at the start of the century.[132] Unlike science of the previous century, which was reserved for the elite men of science and a small portion of the educated upper class, early-nineteenth-century science publications were directed at the emerging social and intellectual niches among the reading public.

As the readership for science became more diverse, so too did its creators. Opportunities for women to cultivate a place for themselves within the elite circles of universities and institutions would remain slim until well into the twentieth century. But the need for more writers combined with the public desire to consume science and natural history created a brief opening through which women could enter the world of science. As science and natural history writing and popularization became a burgeoning literary field unto itself, women found a place for themselves not only as audience and consumers of science but as writers and teachers who shaped public conversations and beliefs about nature and science.

THE MATERNAL TRADITION OF SCIENCE WRITING

One of the earlier women writers of this Victorian trend was Jane Marcet (January 1, 1769–June 28, 1858). Born Jane Haldimand in London on New Year's Day 1769, Jane was the oldest of the twelve children of Anthony Francis Haldiman, a wealthy Swiss banker, and her mother, who was also named Jane. Tutored at home with her siblings, Jane received a high-quality education typical of an upper-class family, and in keeping with eighteenth-century Enlightenment standards of education, she would have received some form of instruction in science (see page 62). Her more serious entry into science came later, after she married fellow Swiss Alexander John Gaspard Marcet in 1799.

As a respected physician, Alexander was well-connected in Britain's intellectual milieu, a network that Jane also benefited from.

It was also through Alexander that Jane turned her focus to chemistry. Alexander's surviving autobiographical memoranda show that he began reading his notes for a chemistry course aloud to Jane in 1801. This was also the year that Humphry Davy began presenting public lectures on chemistry at the

---

With the high output of printed materials and the increase in demand among readers, publishers turned to women writers out of necessity, and women seized the new opportunities open to them.

Royal Society, which were attended by both Jane and Alexander. According to Alexander's notes, Jane had begun writing her *Conversations on Chemistry* in the spring and summer, a book that would become much beloved by readers.[133] Jane shared manuscript drafts with other chemists, who offered commentary and advice on revisions, while her husband fact-checked and assisted with revisions. In most husband–wife scientific marriages the husband usually received credit, but in the case of Jane and Alexander, attribution went to Jane (see pages 47–55).

In 1806, Jane published *Conversations*, which is considered the best-selling English-language book of the first half of the nineteenth century.[134] Michael Faraday credited the third edition with inspiring him to pursue science. He encountered the book while working as the bookbinder's apprentice in the shop where Jane's book was being bound. *Conversations* went through sixteen editions, the last published in 1853, and three French translations. Her name did not appear as author until 1837, and when it did, it appeared as "Mrs. Marcet."

Marcet's target readership for *Conversations* was women, and she used a conversational format to reach them. In the book, the character of Mrs. B.

→

JANE MARCET

Popular science writer
Jane Marcet, author of the
widely popular 1806 book
Conversations on Chemistry,
one of the best-selling English-
language books of the early
nineteenth century.

←

MARGARET GATTY

Margaret Gatty, writer and
periodical editor, wrote *Parables
of Nature*, one of the most popular
children's book in late nineteenth
century, and *British Sea-Weeds*, a
stunningly illustrated two-volume
guide to marine biology.

teaches sisters Emily and Caroline as they work together through various chemical concepts through lively discussion, ultimately arriving at mutual understanding. This format is characteristic of the "maternal tradition" of science writing, which was widely adopted by women in the early nineteenth century. These discussions often took place in a domestic setting and nearly always featured a mother or a specifically maternal teacher like a governess. For modern readers, the domestic sphere might seem like a confining and oppressive space for women, but by giving women characters the expertise of a scientific teacher, they became authorities in science as well as in the more socially accepted feminine domestic sphere. Marcet brought chemistry out of the laboratory and into the home for her female audience, and it clearly worked. Aside from its popularity in Britain, *Conversations* became a standard text in many girls' schools and womens' colleges in the United States.[135] Marcet would use this format for her later books as well, writing *Conversations on Political Economy* in 1816 and *Conversations on Vegetable Physiology* in 1829.

WRITING THE DIVINE INTO SCIENCE

In the latter part of the nineteenth century, the maternal tradition was going out of style and becoming increasingly unpopular for women writers, who wanted to appeal to both male and female readers. One such writer was Margaret Gatty (June 3, 1809–October 4, 1873), who expressed her dislike for Marcet's style in 1864, writing in a letter to friend and colleague, the nineteenth-century naturalist William Harvey, declaring, "I *decline* to read Mrs. Marcet altogether...I believe I hate Mrs. Marcet."[136] Gatty not only used a different format than Marcet to compose her work, she also used a different medium—the periodical as well as the book.

Born Margaret Scott in Essex on 3 June 1809, Gatty was the daughter of clergyman Alexander John Scott, a Royal Navy chaplain, and Mary Frances. Mary Frances died when Gatty was only two, leaving her and her older sister Horatia in the care of their father. When Mary Frances died, the reverend withdrew from family and friends and took refuge in his library, and even though his behavior did not provide young Margaret and Horatia with paternal warmth, he at least provided them with books.[137] Margaret and Horatia received a basic education—piano, drawing, and reading and writing—from a governess, but the real foundation for Margaret's later writing career came from the sisters' own initiatives in writing original poetry and stories. They even started their own writing clubs, the Black Bag Club in 1828 and the Fun Club in 1832.[138]

Margaret, however, laid down her pen after she married the vicar Aflred Gatty in 1839, and together they would have ten children. As a vicar's wife, Gatty was preoccupied with serving and supporting her husband's congregation and community, and as a mother many times over, she had her hands full with domestic duties. But after the birth of her seventh child in 1848, Gatty was sent to the seaside of Hastings to recover from illness. While she recuperated, her physician gave her a copy of William Harvey's *Phycologia Britannica*, a compendium of British seaweeds with beautifully colored illustrations.[139] Gatty was soon enraptured by seaweeds and marine life, and when she returned home, she once again picked up her pen to indulge a new fascination that held her imagination for the rest of her life.

Gatty published her first book, *Parables from Nature*, in 1855, and it became one of the most popular children's books in the latter half of the nineteenth century. Each chapter explores a lesson—either in morality or natural history or both—through anthropomorphized nature: birch trees talk to wood-pigeons, and zoophytes wager with seaweeds. *Parables* was followed by more books including *The Human Face Divine: And Other Tales* in 1860. Perhaps her most notable book was 1862's *British Sea-Weeds*, a stunningly illustrated two-volume guide to local marine botany. It was the culmination of fourteen years of collecting and observing her own specimens of algae and seaweeds. In addition to books, Gatty wrote about natural history for a number of periodicals, including one that she founded titled *Aunt Judy's Magazine*. With a family of ten children, the Gattys often faced money troubles, and with her writing, Gatty was able to provide significant supplemental support for her family. Women like Gatty could remain within the realm of Victorian respectability since their studies and writing could be done from home. Combined with the lessening social stigma surrounding women who earned their own income, Gatty was able to build both reputation and livelihood in science writing.

Running throughout Gatty's work is a marriage between the study of nature and contemplation of God. Theology was often an important component of women's nature writing. Many were influenced by the Anglican William Paley and his 1802 *Natural Theology or Evidences of the Existence and Attributes of the Deity*. Paley argued that nature contained evidence of a supreme being's existence and that new discoveries about nature were proof that the deity was wise, powerful, and purposeful, and that the mere existence of plant and animal life implied an intentional designer. To peer into nature's intricacies was to encounter a divine creator; to study and contemplate nature was to seek divine knowledge. Women were seen by Victorian society as moral vanguards of home and hearth, and teaching morality and piety through a relationship with nature was understood to be well within women's realm of expertise. Paley and the tradition of natural theological writing that these women took up had a profound impact on public writing about natural history and science, particularly when they contended with Charles Darwin and his work at mid-century.[140]

DEBATING AND POPULARIZING DARWINIAN EVOLUTION

The women who wrote in a Paleyan tradition of natural theology would ultimately have to address the theory of evolution as well as the vanguard of scientific authority and secularism ushered in by T.H. Huxley and John Tyndall, among others.[141] Gatty rejected evolution through natural selection and the materialistic worldview it seemed to engender. Publicly responding to Darwin's theory of natural selection, she published the short story "Inferior Animals" in 1861, two years after Darwin's *On the Origin of Species* appeared. The story was a moral tale that questioned the evolutionary argument of hierarchy. As an early commentator on Darwin and a popular writer, Gatty shaped beliefs about Darwin.[142] Even though history and science has not born out her doubt, it is

↑ Engraving depicting the Marioni's rotary press, nineteenth century.
This machine could turn out **7,000** copies per hour with hard packing on dry paper.

significant that she was able to participate in scientific debate at all, when such public platforms and opportunities were so often denied to women in previous centuries.

Arabella Buckley (October 24, 1840–February 9, 1929) also commented on Darwin's theory of natural selection, but unlike Gatty, she did not see spirituality and evolution as mutually exclusive worldviews. Born in Brighton in 1840, Buckley became secretary to the famous geologist Charles Lyell at the age of twenty-four and worked for him for eleven years until his death in 1875. As Lyell's secretary, Buckley was well-connected to Britain's elite men of science, including Darwin and Alfred Russel Wallace, the co-developer of natural selection. Embedded in this intellectual circle, Buckley was well aware of the divisions between Darwin and his contemporaries, many of whom could not reconcile the evolution of humans' so-called "higher" instincts like morality with a system of natural selection that did not require a creator or divine intervention.

As multiple cultural changes converged in nineteenth century Britain, a new opening into science and public life briefly presented itself and women seized it without hesitation.

In his 1871 book *The Descent of Man*, Darwin addressed his contemporaries' concerns and explained human morality as the result of filial affection, a key trait that increases the survival of offspring developed through generations of natural selection. Buckley found this argument compelling, and she defended it in her first publication in *Macmillan's Magazine*, titled "Darwinism and Religion," just three months after the publication of *The Descent of Man*. Buckley affirmed that natural selection did not jeopardize belief in God and the immortality of the soul, nor did it diminish the superiority of human consciousness. She argued that filial affection for survival had evolved into an imperative to take care of one's community, an instinct that privileged self-sacrifice, not self-interest. Writing in the same year as *The Descent of Man* appeared, Buckley was the first writer to advocate for a mutualistic reading of Darwin.[143]

Buckley would take up the topic of evolution again in the children's book *Life and Her Children* in 1881 and *Winners in Life's Race* in 1883. Across the two books, Buckley wove a highly imaginative evolutionary epic as she guided readers through the various forms of life, starting with the lower life forms and ending with humans in *Winners*. Not only did she illustrate the basics of natural selection, she further developed the view of natural selection that she advocated in *Macmillian's* a decade before. Her evolutionary writing showed that species survive through co-operation and self-sacrifice to the community. Buckley was one of the most influential popularizers of Darwin's natural selection. Throughout her writing career, she maintained correspondences and even friendship with male scientists, like Darwin and Wallace, even as she developed her innovative theories about natural selection on the periphery of science as one of the most authoritative science popularizers of the century. Between 1876 and 1901, Buckley published ten books on science, including her most creative works *The Fairy-Land of Science* and its sequel *Through Magic Glasses*.[144] In both, she used language and imagery from fairy tales to overlay the natural world with a sense of wonder and magic.

The work of nineteenth-century women science writers and popularizers does not conform to the model of scientific achievement that privileges theoretical science, invention, and discovery. Such practices and institutions were typically reserved for the likes of Charles Darwin or Michael Faraday. Instead, women carved out a space for themselves beyond the boundaries of the institutions, universities, and societies that would not have them. As multiple cultural changes converged in nineteenth-century Britain, a new opening into science and public life briefly presented itself and women seized it without hesitation.

Pl. 28

1.Belladonna purpureis. 2.Belladonna blanda. 3.Belladonna purpureis pallida.

Day & Haghe Lith.ᵗˢ to the Queen

CHAPTER 8

Botany for Ladies

Victorian England's fascination with natural history can be aptly described as an obsession. Words like "mania," "fever," and "craze" characterize popular natural history trends in the nineteenth century and paint a vivid picture of the intensity with which people consumed knowledge and images of the natural world. The century was marked by gardening crazes and fern-fever; fervent insect-collecting and butterfly-mania; aquarium fever; and fossil-hunting and rock-collecting fads.[145] All these obsessions echoed their more formally scientific partners: botany, entomology, conchology and marine biology, and geology. But to participate in these crazes, one did not need to be specially trained or educated in the associated science or be elected to a scientific society—one simply needed the enthusiasm.

People could now also feed their natural history cravings with the hundreds of natural history books and periodicals that flew off the printing presses.[146] Natural history columns became a common and expected feature in newspapers, and naturalists and science popularizers fashioned new writing careers dedicated to the various crazes. As the popularity of natural history grew, women's writing on science and nature also flourished. While theoretical science remained within the domain of gentlemanly men of science, natural history crazes became an increasingly female sphere.[147]

Natural history encouraged observation of the particular in nature over abstract theory because it was in the particulars of nature—the tiny branching veins of a leaf, the microscopic sensory hairs of a day-fly, or the unrivalled variety in colors of insects —where beauty, and evidence of a divine creator,

← Lithograph of a Belladonna Lily: *Amaryllis belladonna* from Jane Loudon's *The Ladies' Flower Garden of Ornamental Bulbous Plants*, published in London in 1841.

could be found. The egalitarian nature of natural history also appealed to women as collectors, observers, writers, and readers. It gave women room to explore nature, collect its curiosities, observe its intricacies, and ultimately to write about it on their own, despite exclusion from higher learning and societies. The Victorian passion for these crazes ensured a steady demand and sustained readership for the work that women had to offer.

BOTANY AS A PROPER SCIENCE FOR LADIES

Women published most prolifically on botany and its sibling subjects, gardening and horticulture. Botany had long been the domain of women as midwives, healers, and caretakers, who used plants for medicinal purposes. For eighteenth- and nineteenth-century women writers and readers, botany was both a source of spiritual edification and a way to cultivate womanhood through a relationship with nature. More than any other science, botany aligned with cultural beliefs about women's nature and their supposedly natural province as mothers and wives.[148] Well into the nineteenth century, botany was considered more feminine than other sciences and thus appropriate and "proper" for women. Through botany women found more outlets through which to indulge their botanical enthusiasm, since they comprised a large portion of the market for the botany popularizations that fed fern-fevers and collecting crazes.

Just before the turn of the century in 1796 Priscilla Wakefield (January 31, 1751–September 12, 1832) published *An Introduction to Botany, in a Series of Familiar Letters* in 1796, the first systematic introduction to the science written by a woman. Composed as series of letters written between two teenage sisters, the book introduced readers to the Linnaean system of botanical classification as one sister relays her governess's lectures on botany to the other. Swedish botanist Carl Linnaeus published his *Systema Naturae* in 1735, and despite there being as many as fifty-five systems of classification in 1799, the Linnaean system was the most favored in the late eighteenth and early nineteenth century.[149]

Linnaeus argued that the reproductive organs of the plant were the most important part of plant biology for classification purposes, so he organized his entire system around the male stamen and the female pistil sex organs. The number of stamens in a plant determined the "class" to which it belonged while the number of pistils determined the plant's "order." This was however an artificial system with no grounding in nature since the sex organs of the plant in and of themselves do not indicate any hereditary or phylogenic pattern.[150] But the Linnaean system was simple to understand, enabled easy classification through

↑

PRISCILLA WAKEFIELD

Priscilla Wakefield published the first systematic introduction
to botany written by a woman, *An Introduction to Botany,*
in a Series of Familiar Letters.

observation of a plant's visible features, and was fairly consistent across a wide variety of species.[151] Its simplicity was not the only reason women preferred it for introductions to the science; it was easily adaptable to women's style of writing at the time. Once the plants were separated based on their male and female sexual anatomy, it was a simple next step for writers to gender the plants and impose nineteenth century masculine and feminine gender roles onto them.

Even though this type of gender stereotyping seems regressive and even oppressive given women's relegation to the domestic sphere during this time, it made room for ways of writing about science that embraced sentimentality, romance, and femininity. These characteristics of women's science writing conformed to cultural expectations for women writers and ironically gave them a unique authority to write about botany. For instance, Frances Rowden (1774–circa 1820–1840) composed *A Poetical Introduction to the Study of Botany* in 1801, a lyrical book on Linnaean botany for children and women that taught lessons in ideal womanhood. When Rowden explained the classification of lily of the valley, she glorified female modesty and chastity: lily of the valley has one female pistil to six male stamens, and for Rowden, the flower becomes a personification of virginal modesty, a maiden who maintains her innocence while surrounded by all the male attention of their stamens.[152] Women writers similarly appropriated the Linnaean system so frequently that the Linnaean system became seen as a feminized way to do botany.

TAKING THE FEMININE OUT OF BOTANY

Around the 1830s, men of science began attempting to professionalize the science and rescue it from its amateur status and connection to natural history popularization. In attempting to bring botany to the sphere of academic and professional science, an arena to which women still had little access, men of science sought to regain control of what had become a thoroughly feminized field. Botanist John Lindley was particularly vocal and public about wresting botany from the pens of women, proclaiming in his inaugural lecture at the University of London in 1829, "It has been very much the fashion of late years, in this country, to undervalue the importance of this science, and to consider it an amusement for ladies rather than an occupation for the serious thoughts of men."

Over the next several decades, botany would split between feminine amusement and masculine seriousness in a number of ways, reflecting Lindley's own distinction between the two.[153] First, the natural system of classification, championed by Lindley, gradually usurped the Linnaean system. Where

Linnaeus's system provided an easy, though incomplete, means of plant identification, the natural system took into account patterns of heredity among plants and their phylogenic relationships when classifying them. What was more, Lindley preferred the natural system because it did not have the same association with women and femininity as the Linnaean system did. As a result of these

For eighteenth- and nineteenth-century women writers and readers, botany was both a source of spiritual edification and a way to cultivate womanhood through a relationship with nature.

changes, the language of botany was split between the literary and narrative forms of botanical writing in the amateur and popular realm, and a more specialized and de-personalized style of writing for the professionalized science. Where once there was one, feminine form of botany, it had now been divided into an amateur "lady science" and a male-dominated professional one.[154]

The popular writings of women like Wakefield and Rowden would have no place in this new future for botany. By the 1830s, women's traditions of writing about botany—family narratives, maternal storytelling, dialogues, and letters—began to fade out and by the 1850s they had mostly disappeared.[155] Women who wanted to continue writing about and practicing botany would have to adjust to the more specialized form and adapt a "scientific" way of writing about it. This necessitated a change in who they were writing for.

In the course of her career, botany popularizer Jane Loudon (August 19, 1807–July 13, 1858) witnessed this shift and found herself making choices about what to do with her work. Loudon was a prolific botanical writer, publishing five books on botany and gardening and holding two editorships at women's magazines, *Ladies' Garden Magazine* and *The Ladies' Companion*. Her most popular books include *Botany for Ladies* published in 1840 and *The Ladies' Companion to the Flower Garden* a year later. Since the Linnaean system had been

↑

LYDIA BECKER

Oil painting of Lydia Becker by Susan Isabel Dacre.
Becker was an editor, science writer, and women's rights advocate.

overturned by the time she published *Botany for Ladies*, Loudon adopted the natural system of classification, but she still wrote these books with an audience of women in mind. But a decade later in 1851, the title of the second edition of *Botany For Ladies* was changed to *Modern Botany*, removing women as the preferred audience of the book.[156]

Lydia Becker (February 24, 1827–July 18, 1890), another botanical writer, made similar choices with her work when she published *Botany for Novices: A Short Outline of the Natural System of Classification of Plants* in 1887. Like Loudon, Becker intended to write an introduction to the natural system for young women, but nothing in the title or book itself makes that clear. Unlike with women's traditional writing about botany, which always featured women, nothing explicitly indicates the gender of the intended audience or the author, since Becker published under her initials. An ardent advocate for women's rights and equal education, Becker argued in front of the British Association for the Advancement of Science in 1868 that the mind has no sex, saying that "there is not distinction between the intellects of men and women corresponding to, and dependent on, the special organization of their bodies."[157] Becker likely saw her choice to remove any indication of gender as a move toward a gender-neutral science, one that obscured distinctions based on sex. But as Lindley had made clear, this was a male model of science and science writing that distanced itself from domestic culture, sentimentality, romance, and morals—all of which were believed to be the realm of women. The diminishing marketplace for botanical books specifically for women served to erase them from their place in the science.[158]

CHANGING THE NARRATIVE

Although the place of women in botany was much changed by the end of the century, Elizabeth Clarke Wolstenholme Elmy (see page 96) was an exception to the new rule of masculine science writing. Elmy published two botanical texts for children that used earlier women's writing traditions: *Baby Buds* in 1895 and *The Human Flower: A Simple Statement of the Physiology of Birth and the Relations of the Sexes* in 1896. Curiously, Elmy was not a science popularizer or a botanist, but was first and foremost a feminist and organizer for women's education and liberation, and she enlisted botany in her feminist cause.

Elmy was particularly concerned with women's sexual liberation, especially women's sexual consent and the repeal of the Contagious Diseases Acts (CDAs). Elmy was the first person to publicly call for an end to marital rape, which was legal at the time, and to argue for its criminality.[159] Her exploration of sexual consent was connected to her activism against the CDAs, a series of public health measures that attempted to stem the spread of venereal disease throughout Britain. The CDAs disproportionately targeted women by giving police authority to forcefully detain and examine any woman believed to be a sex worker while their male clients escaped such punishment and continued to spread disease unhindered. The gendered double-standard in British culture that criminalized female sexuality while male sexuality remained unrestricted threaded together both issues of concern for Elmy. Her botanical texts were an extension of her fight against this double-standard.[160]

Originally published under the pseudonym Ellis Elthemer, *Baby Buds* was a sex education primer for children that passed as an introduction to botany, and *The Human Flower* served the same function but for young adults. *Baby Buds* featured a mother-narrator who uses the sexual reproduction of plants to introduce children to human sexual anatomy, intercourse, and relationships. The narrator teaches the child not only about the act of sex and reproduction

↑ Political cartoon by John Tenniel in the 28 May 1870 issue of *Punch* depicting Lydia Becker leading a group of women demanding John Bull, a fictional personification of England, for their right to vote.

but about sexual responsibility. In walking the child through intercourse to conception to birth, the narrator shows that both men and women contribute equally to sex and to reproduction.[161] Elmy also encouraged sexual consent, writing that humans do not need bees to transfer their pollen because the male

---

Where once there was one, feminine form of botany, it had now been divided into an amateur "lady science" and a male-dominated professional one.

---

and female "approach *one another* [emphasis added] for that purpose".[162] In the late nineteenth century, addressing the cultural double-standard in male and female sexuality, especially to children, was a radical act. By returning to women's writing traditions at the end of the century, long after these traditions had disappeared from the literary marketplace, Elmy was choosing her audience carefully. By speaking directly to children and people who were uneducated about sexual reproduction, she attempted to shift cultural beliefs about male and female sexuality. Early women's writing traditions on botany that lent themselves easily to human analogy and educational lessons were more suited to Elmy's needs than the more modern scientific way of approaching the subject. Elmy's significance lies in how she harnessed the science in service of her larger activism for women's liberation.

Women's writing about botany in the nineteenth century was incredibly diverse in both the forms of their narratives and in the agendas they advocated. At the beginning of the century, Frances Rowden used botany to instill in young women the virtue of modesty and chastity while at the end of the century Elizabeth Elmy used botany to educate children about sex. Even though the women advocated opposing agendas, the science of botany and the market for science popularizations gave them a public platform from which they could be heard. But when men of science sought to claim botany as theirs—to confine it to the realm of "the serious thoughts of men"—women effectively lost that public platform and the entry into the science that it had allowed them to have.

Elizabeth Clarke Wolstenholme Elmy

(baptized December 15, 1833–March 12, 1918)

Born in 1833 in Manchester, Elmy was the daughter of a Methodist minister and a working-class mother, but by the age of twelve, she was orphaned. Her maternal grandfather, Richard Clarke, took charge of her care, and when he died, Clarke left her enough money to attend school at Fulneck Moravian School. She went on to become the headmistress of a private girls' school and founded the Manchester Schoolmistresses Association in 1865.

By the 1850s, Elmy had committed herself wholeheartedly to a career rooted in women's emancipation and enfranchisement, participating in more than twenty feminist organizations. Elmy's seemingly arbitrary entry into botany at the age of sixty-five was an extension her activism to repeal the Contagious Diseases Acts.[163]

Elmy's botanical primer *Baby Buds*, which introduced children to human sexuality through the sexual reproduction of plants, was published in 1895 under the pseudonym Ellis Ethelmer. She quickly followed *Baby Buds* with another botanical sex-education book for young women, *The Human Flower: A Simple Statement of the Physiology of Birth and the Relations of the Sexes*, in which she appropriates the same themes for a more mature audience.

Elmy died in 1918. Her contemporaries saw her as one of the most influential voices in the struggle for women's freedom. Her significance in botany, however, is less known because she did not make any new or innovative discoveries.

CHAPTER 9

From the Home to the Hospital

BECOMING A NURSE IN AMERICA

In early nineteenth-century America, nursing was not yet the profession we recognize today with its specialized training schools, professional societies, and defined duties of care. After the American Civil War, nursing began to take shape as a formalized and legitimate field of medicine. But nursing's roots stretch back much farther to the caring that has almost exclusively been done by women in the home. Unlike most of the fields described in this book, nursing was always dominated by women because the very foundation of nursing was based on strictly gendered ideas about women and the type of work to which they were supposedly most suited. This close association between women's domestic labor and nursing would profoundly influence the field as it professionalized throughout the nineteenth century. American nursing was shaped by the Civil War in ways that set it apart from nursing in other parts of the world. Even as American nurses adapted ideas and methods from abroad, they built a unique profession that was suited to the extraordinary pressures of American life in the nineteenth and early twentieth centuries.

The work of caring for the sick and injured that we typically associate with nursing today was originally performed by women for their own families and communities. As with the midwives and women medical practitioners of pre- and early-modern periods, providing medical care for one's family members was a task almost every woman could expect to do in her lifetime.[164] Although caring for the sick was among the domestic duties of wives and daughters, it was not unskilled labor. Many women educated themselves about medical

← Nurses leaving the Henry Street Settlement. The settlement house was founded by Lillian Wald in 1839 to serve the poor residents of the Lower East Side in New York.

matters by attending lectures or reading handbooks and pamphlets.[165] Equally important was experience and wisdom passed down through generations of women before them. Women were not paid for their medical care and expertise, and they would not be until well into the nineteenth century. Nor were women providing this care in an institutional setting. Indeed, the hospital itself was not the centralized location for medical care that we think of today, and most people received medical care at home, provided by women family members, neighbors, or, if the patient could afford it, physicians. The authority and knowledge of women who provided nursing care in this context was often in direct opposition to that of physicians, whom nurses saw as encroaching on their long-held territory.[166]

American hospitals in the early nineteenth century were sometimes maintained publicly but often by voluntary and reform organizations, and the care they provided was oriented toward chronic illnesses and poverty, rather than acute illness or trauma. Nurses in these hospitals were often the more ambulatory women patients themselves, and their jobs were more closely associated with domestic labor like cleaning and laundry than providing medical care.[167] In the mid-nineteenth century, as industrialization and urbanization profoundly changed the American economy, women were able to transfer their caring experience into paid employment by offering nursing services in their communities.[168] Like the medical "grannies" of Imperial China (see pages 22–23), most of these women were older, and many had been forced into working outside the home by the deaths of their husbands.[169] Many of these women worked in the homes of community members, caring for the sick, rather than in the hospital setting we now see as the nurse's domain.

NURSING DURING THE CIVIL WAR

Nursing in America was not just shaped by stringent gender roles for women but also by slavery, and later by the war itself. Many people who were enslaved and brought to America had extensive healing knowledge from their own cultures, which enslaved women often harnessed to care for other enslaved people and for their white owners' families.[170] Enslaved women who were trained as midwives were additionally valuable to slaveowners, who could rent their services to other families.[171] The cultural image of caring labor among enslaved women in the American South has been dominated by the figure of the "Mammy," or wet nurse, who cared for the slaveholder's children and managed the household.[172] In the North, especially during the Civil War, Black women worked in military hospitals and as nurses and midwives in free cities.[173]

The title "nurse" did not always refer to a person who cared for the sick or injured. Instead, it often referred to the practice of wet nursing and infant nursing, which was largely taken up by enslaved people or hired domestic laborers who breastfed, cared for, and raised to adolescence the babies of the families they were owned or employed by. In an 1861 edition of *The New York Times*, a page of classified advertisements listed some forty advertisements for nurses, only two of which sought a nurse to care for an "invalid lady."[174] The majority of the advertisements asked for an applicant who "can take care of a baby from its birth," and "bring it up by hand," as well as a variety of other domestic duties, usually sewing. The ads specified nurses and/or seamstresses would be working for private families, and it was clear that some would be expected to accompany the family to their country homes outside the city as well.[175]

As the American Civil War strained the existing medical infrastructure, the need for a trained nursing workforce became increasingly urgent. Desperate to help the war effort, women with no formal training but years of experience caring for their own families flooded hospitals offering their services as nurses. Nearly 20,000 women of various backgrounds volunteered as nurses during the

↑ Army nurses of the Spanish American War in Chickamauga, Georgia in 1898.

war. But they were swiftly sorted by race and class into duties that matrons and administrators felt fitted their social stations: white women served as matrons and nurses, and Black women were relegated to service as cooks and laundresses.[176]

After the Civil War, nursing was slowly professionalized through a combination of experiences from the war and women's organizing efforts, which were motivated in part by a need to train the daughters of families disrupted by the war for the workplace.[177] Using the organizational expertise gained by women who worked in medicine during the war, nursing schools began training the women who would be the first "graduate" nurses in the United States's history in the early 1870s.

THE NIGHTINGALE MODEL AND U.S. NURSING

Early professional nursing in the United States was heavily influenced by Florence Nightingale (May 12, 1820–August 13, 1910), whose 1859 book *Notes on Nursing: What it is, and what it is not*, first appeared in an American edition in 1860.[178] Although Nightingale's wide acclaim was a result of her innovative practices as a nurse during the Crimean War, she also understood that "every woman is a nurse," in that women in her era were expected to have "charge of the personal health of somebody" at some point in their lives. In her approach to nursing, Nightingale prioritized the creation of an environment that promoted healing. She had observed that the symptoms of many diseases were a result of patients being treated and recuperating in less than ideal environments. Lack of fresh air or proper temperature control as well as rigorous scheduling of care and treatment, compounded by a dirty hospital ward or home sick room, were often more important factors in patients recovering their health than medicine or surgery.[179]

Even though nursing was willing to adopt some of Nightingale's medical approaches, other women leaders in medicine in the nineteenth century were at odds with Nightingale's gendered philosophy of nursing. In her view, the qualities of a good nurse were inherent to exemplary womanhood itself; the woman who wished to become a *doctor* instead was abandoning the specific virtue that called women to care for the sick. As historian Patricia D'Antonio writes,

> "[n]ursing knowledge was distinct from medical knowledge: it was 'everyday sanitary knowledge' of fresh air, warmth, proper ventilation, nutritious diets, exemplary hygiene, and calm environments that all women [...] needed to assume responsibility for learning and for wielding."[180]

Nightingale's influence was wide-reaching, but not all nursing education in the nineteenth century followed the Nightingale model. As early as 1839, physicians began developing curricula for nursing that emphasized scientific knowledge of the body and medical therapies. In Philadelphia, a male physician developed a nursing education program in part as a way to train a workforce separate from midwives, who were responsible only to the women whom they assisted in childbirth. These nurses would help to bolster the scientific authority of the physician.[181] As physicians became more committed to the scientific principles of observation and objectivity in their healing work, they recognized the need for trained nurses to help them gather and process this new data about patients and to create new innovative treatments.[182] This role for nurses, as lieutenants within the physician's highly trained hierarchy, came with a new interpretation of the women already working as nurses at home or in hospitals, who were now seen as woefully unqualified to aid in the mission of scientific medicine.[183]

THE BEGINNING OF RACIAL INTEGRATION IN NURSING

Just as Civil War nursing was segregated by race and class, professional nursing followed a similar pattern. The image of Nightingale as "the Lady with the Lamp" was profoundly gendered but also predominately white. Yet, Black women still excelled in professional nursing despite the racial barriers within the field and post-Civil War society at large. Mary Eliza Mahoney (May 7, 1845–January 4, 1926), a Black woman who trained at the New England Hospital for Women and Children, became the first African American woman to receive a formal nursing degree in the United States in 1879. Her entry into the profession through formalized training marked an important moment in the history of nursing, when schools began to racially integrate training.

Born in Boston, Massachusetts in 1845, Mahoney lived in the free North, but still grew up in a deeply segregated environment in which fugitive slave laws were in effect and Black families routinely faced housing and employment discrimination and racist violence.[184] Like white women, Black women worked as nurses in private homes, but their roles were conditioned by racial labor hierarchies that made it more difficult for Black women to separate nursing duties from other domestic labor. The new kinds of authority that white women gained in their associations with physicians and nursing training were largely unavailable to Black women.[185] Beginning work as a nurse in people's homes in 1865, Mahoney was constrained by these racialized dynamics.

Around the same time, however, women advancing the cause of women in medicine founded the New England Hospital for Women and Children in Boston with the goal of training women physicians and nurses and providing medical care to women by women.[186] Unlike other segregated public and voluntary hospitals at the time, this hospital admitted patients from all backgrounds. In 1878, six years after the hospital had relocated to Mahoney's home in Roxbury, she was admitted to the sixteen-month nurse training program after working for a time in the hospital cooking, cleaning, and washing.[187]

By the time Mahoney started training, the program was more than a decade old and highly rigorous. Students worked sixteen-hour days, supervising their own wards, attending lectures on topics including appropriate food for sick people, techniques of surgical nursing, and nursing within private families. Students rotated through the different hospital departments, gaining clinical experience in each.[188] Mahoney graduated in 1879, and the hospital continued to admit Black students after her, training five other Black nurses over the next twenty years.[189]

Now credentialled and wielding the authority of her training, Mahoney returned to nursing in private homes. Records of Mahoney's success as a private duty nurse remain because she registered with a nursing directory maintained by the Boston Medical Library, which preserved her references for potential employers. Previous employers recommended her as an "excellent nurse."[190] Even as an "excellent nurse," Mahoney and other Black women working in the homes of white people still faced discrimination and indignity. In private homes, Black workers were never allowed to eat in the same space as their white employers, even when white nurses could often do so. Mahoney was forced to eat in the family's kitchen, but she insisted that she be allowed to do so at times when the other servants were not eating.[191] This anecdote from Mahoney's work as a highly trained, respected nurse encapsulates many of the conflicts and contradictions of nursing in the nineteenth century.

NURSING IN THE PROGRESSIVE ERA

The new identity of the formally trained nurse as an intellectual worker rather than a domestic servant was catching on, even as many in the medical community continued to harbor deep suspicions about investing women with medical knowledge. In 1890, there were not even 500 trained nurses in the whole country, but by the turn of the century there were an estimated 3,500.[192] In the early twentieth century, nursing was still undergoing deep professional, social,

←

Nurses say goodbye to
a recovering patient,
World War I.

→

LILIAN WALD

A towering figure in the history of public health in
America, Lilian Wald founded the Henry Street Visiting
Nurse Service in 1895 in New York City to care for
people living in poverty on the Lower East Side.

←

MARY MAHONEY

In 1879, Mary Eliza Mahoney became
the first African American woman to
graduate with a professional nursing
degree in the United States, and went on
to have a successful career as a private
nurse. Her success opened the way for
Black women to train as nurses.

and scientific change. Nursing in hospitals and private homes became more professionalized, and training of nurses became more rigorous and standardized. Nursing, reflecting its origins in voluntary associations and the charitable work of wealthier women, also became an important force for public health reforms.

From the caring traditions of enslaved African women to the ambitious social reform programs of middle-class women, the diverse experiences of the women who became trained nurses in the nineteenth century profoundly shaped the profession that we know today.

These shifts in nursing were part of what historians call the Progressive Era, a period of wide-ranging social and political movements and reforms in the United States in the late nineteenth and early twentieth centuries.

Progressive Era reforms in nursing and public health, as in other areas, were often led by women. In 1889, a young German-Jewish woman named Lillian Wald (March 10, 1867–September 1, 1940) entered training at the New York Hospital Training School for Nurses. Wald came from a middle-class family and grew up in Rochester, New York. She cited her experience with a nurse who attended her older sister in childbirth as her inspiration for becoming a nurse, but as historian Marjorie Feld has shown, it is clear that Wald's ambitions for her life and work were always much larger.[193] After leaving nursing school in 1891, she went to work at the Juvenile Asylum in New York City, but quickly became disillusioned with the institution and returned to school at the Women's Medical College in Manhattan for a time.[194]

As a middle-class woman, Wald had many connections to other wealthy and reform-minded women, whom she partnered with in her work. In 1893, she and fellow nursing school graduate Mary Brewster moved to the Lower

East Side, a section of New York City where many immigrants lived and where poverty and overcrowding had become dire by the turn of the century. In 1895, Wald and Brewster established the Henry Street Visiting Nurse Service after living in a tenement and learning about the community they hoped to serve. This type of nursing service, where nurses lived in the communities they served and provided education and medical care, was called a settlement house. In her memoir, Wald wrote that she understood her training as a nurse to be valuable to helping the poor. After visiting a sick woman on the Lower East Side and seeing the family's circumstances, Wald "...rejoiced that I had had a training in the care of the sick that in itself would give me an organic relationship to the neighborhood in which this awakening had come."[195]

Wald and Brewster hired nurses as well as lawyers, union organizers, and other social reformers.[196] Nurses were dispatched to the community from Henry Street and used a sliding scale for fees to ensure that everyone had access to medical care. It was Wald who coined the term "public health nursing" to describe this type of system, which also included the provision of educational classes.[197] As was the case earlier in the nineteenth century, most people did not go to a hospital when they were sick. Wald was sensitive not only to the sometimes prohibitive cost of hospital care but also the social strain of women in particular leaving the home for care.[198] Wald also wrote in her memoirs about the times she and the other nurses had to actively subvert the authority of doctors to see that their patients receive proper care.[199]

Mahoney and Wald were both unusual presences in the predominantly white, Christian world of nursing. Wald not only subverted the traditional expectations of her middle-class upbringing by working outside the home with such ambition, she was also a lesbian and made her life within a community of women friends and intimate partners, rather than marrying and having children.[200] The history of nursing in the United States is a complex one, in which forces as large as the Civil War and as small as the personal lives of individual nurses all matter. And although the reach of Florence Nightingale's ideas about the profession was broad, it was not absolute. From the caring traditions of enslaved African women to the ambitious social reform programs of middle-class women, the diverse experiences of the women who became trained nurses in the nineteenth century profoundly shaped the profession that we know today.

FIRST WOMAN'S MEDICAL COLLEGE BUILDING.

AS IT APPEARED AT THE FIRST COMMENCEMENT IN 1850, LOCATED AT 229 (OL NUMBER) ARCH STREET, BELOW SEVENTH.

CHAPTER 10

Home Physicians and Lady Doctors

The practice of medicine has only been a regulated, standardized profession for a relatively short period of time—formal training in medical schools, with attendant credentialing and licensing for physicians, men and women, is a modern phenomenon. As with the profession of nursing, the nineteenth century was a formative time for medicine: the hospital became the central locus of care and medical practice became more regulated and formalized. Interestingly, however, it was sometimes a lack of regulation that allowed women to become physicians since rules restricting their study of medicine in college and universities were not yet in place. The result was a kind of massive professional migration of women seeking medical education in the nineteenth century.[201] In many cases, women from all over the world traveled to the United States to study at one of the many women's medical schools established there. In others, they traversed Europe in search of better educational opportunities. In the United States, the establishment of private institutions for women's education created more opportunities, both for American women to pursue medical education and for women from other countries to seek out sponsorships to travel there to study. In Europe, women seeking medical education faced centuries of tradition that excluded women from state universities.[202]

In nineteenth-century Germany, medicine was already heavily regulated by the state. Doctors could not practice in Germany unless they had been trained by one of the state-regulated colleges, and none of these schools admitted

← The Woman's Medical College of Pennsylvania trained some of the most well-known women doctors of the nineteenth and early twentieth centuries, including many immigrant women who came to the United States specifically to study medicine.

women. Therefore, by law, German women could not hold the title of physician. This was the position that Anna Fischer-Dückelmann found herself in when she decided to attend medical school. Dückelmann came from a middle-class family of physicians, and in her own practice, she would prioritize the health of women and children and the implementation of natural remedies including water cures and vegetarian diets to promote health.[203]

As in other places around the world, German reform-minded women had begun to advocate for better education for women in general and for their access to medical education in particular. German primary education was structured by gendered concepts of the disparate life roles of men and women. Boys were educated according to the principles of the *bildung*, which prepared them both for university and civic life and focused on the cultivation of the self. Girls' education focused on *bestimmung*, which prioritized feminine duty to home and family. Education outside of this duty was considered superfluous, and in extreme cases, damaging to a girl's future reproductive capabilities.[204] Even if women had been allowed to enter universities, the strictly gendered curriculum of German primary education meant that women would have been woefully underprepared to take on university courses. Dückelmann supplemented her own education with preparation in mathematics and Latin, which she studied with her husband.[205]

In 1885, approaching her thirties and married with children, Dückelmann decided the limits on her ability to practice medicine in Germany had to be overcome, and she made plans to enter training at the University of Zurich in Switzerland. The opening of the medical faculty to women at Zurich in the late 1860s brought an influx of women from other countries seeking medical education. For German women especially, Zurich was an ideal choice because the university offered instruction in German.[206] During her education and clinical experience in Zurich, Dückelmann still took care of her three children, including their education, while managing her own studies.[207] She acknowledged the difficulty of doing this, writing that a person with a family to care for ought not pursue medical school "unless she were guided by a specific task and had great willpower."[208] Dückelmann graduated with a medical degree in obstetrics and gynecology in 1896 and moved to Dresden-Loschwitz where she opened a clinic for women and children that remained open until 1914.[209]

Dückelmann understood her duty to her family as an integral part of her own womanhood, but she did not see her duty as a woman and a physician as mutually exclusive. She united these two seemingly opposing roles for women in the figure of the *hausärtzin* or home physician, which she formalized in her popular medical text *Die Frau als Hausärztin* (*The Woman as Home Physician*), which was 1,000 pages in length and contained nearly 500 original illustrations of the anatomy of the body. Integrated into her philosophy of the woman as home physician was her belief that women patients were best served by women physicians. During her time in medical school, she witnessed women, particularly poor women, being treated particularly badly by male physicians and medical students. To counteract these abuses, she sought to provide women with education about their own bodies in order to treat themselves and their families, or know when they were being abused by male physicians. Dückelmann wrote several medical texts over the course of her career, though *Die Frau als Hausärztin* was her most popular; it was translated into several languages and remained in print until 1981.[210] Even though Dückelmann did not instigate sweeping institutional reform to the German medical establishment, she was a critical voice in shining a light on the flaws of male physicians and in elevating the position of women physicians and women's medical knowledge in the home.

WOMEN'S MEDICAL EDUCATION

Perhaps the most famous woman to practice medicine as a physician in the nineteenth century was Elizabeth Blackwell (February 3, 1821–May 31, 1910). Born in England in 1821, Blackwell and her family emigrated to America in 1832. As progressive Quakers, the Blackwells were involved in social causes ranging from the abolition of slavery to women's suffrage.[211] The story of Blackwell's admission to the Geneva Medical College in New York has become a parable about the indignities that faced women seeking medical education. Unwilling to shoulder the full responsibility of rejecting Blackwell's application, the faculty put the question to the student body. Not realizing that the faculty had already decided to reject Blackwell, they sarcastically approved the application with much laughter and catcalling.[212] When Blackwell arrived to study some weeks later, the joke seemed to be on the other students, until the school barred any other women from attending shortly after her graduation in 1849.[213]

Blackwell was the first woman to receive a medical degree in America, and her accomplishment was all the more rare because it was coeducational. Blackwell's hard-won success in a coeducational institution was the original goal for the reformers who led the charge for women's medical education in the second half of the nineteenth century. Initially, they were not focused on creating women's schools because they viewed that option as an admission of defeat in their efforts to integrate women into male medical schools. Advocates for women's medical education, many of whom were physicians like Blackwell,

ANNA FISCHER-DÜCKELMANN

A physician and health educator, Anna Fischer-Dückelmann was a German doctor who pioneered a philosophy of home medicine designed to best serve women and children.

ELIZABETH BLACKWELL

Elizabeth Blackwell, perhaps the most famous lady doctor of the modern era, founded the New York Infirmary for Women and Children and worked with her sister Emily to advocate for women's medical education.

understood that coeducation meant that women would have the exact same credentials as men, which was essential to protecting their reputation as physicians.[214] But Blackwell's educational experience was unique, and medical schools' continued refusal to admit women eventually forced the creation of women's medical colleges all over the country. It was in these colleges that both Americans and women from other countries were able to realize their aspirations to become doctors.

In 1868, Elizabeth Blackwell and her sister Emily opened a medical college for women that was attached to the New York Infirmary for Women and Children that Elizabeth had founded a decade earlier. A number of women who studied or took clinical experience at the Blackwells' college went on to found their own medical institutions, like Marie Zakrzewska (September 6, 1829–May 12, 1902), who founded the New England Hospital for Women and Children. New institutions for medical education for women were also founded by religious groups. A group of Quakers, who were at the time searching for educational opportunities for a number of women, founded the Woman's Medical College of Pennsylvania (WMCP) in Philadelphia in 1850.[215]

It was at the WMCP that some of the most famous women doctors were educated, many of them having traveled long distances from places like India, Japan, the Philippines, Syria, and Russia.[216] The WMCP's roots as a Quaker organization are an important part of the college's mission to educate women from other countries. Many women from colonized countries who sought medical education were sponsored or supported by missionaries. In America in particular, "medical missionaries" were trained at women's medical colleges such as WMCP and many of them committed to continuing missionary work when they returned to their home countries to practice medicine—those that did were offered reduced tuition at WMCP.[217] While these women were undoubtedly pioneers in their quest to become doctors, their education and medical practice were also caught in the current of colonialism and American imperialism that operated in many places around the world. For reformist religious organizations conducting missionary work in colonized places, educating colonized people was a way of promoting the benefits of Western religion and science, often at the expense of Indigenous medical traditions.[218] Some women swam with that current, and others resisted its pull.

INTERNATIONAL MEDICAL STUDENTS

It was Protestant missionaries who helped Anandibai Gopalrao Joshee (see page 121) enter medical school in Philadelphia, after making the long journey to America from India. Joshee grew up in a Brahmin family near what is now Mumbai, and her father, going against Hindu custom, enrolled her in school as a child, hiring a relative named Gopalrao Joshee to tutor her. At the age of twelve, Anandibai was married to Gopalrao.[219]

With Gopalrao's support, Joshee was interested in pursuing her education, particularly in medicine, but it was socially stigmatizing for her to do so within Hindu culture. In her letters, Joshee recalled the experience of being harassed and laughed at for walking in the street with her husband. Gopalrao knew some American missionaries who were working in the community and appealed to them for help sending Joshee to America for medical school, but it was Joshee's own relationships with missionaries that ultimately proved fruitful. A woman in New Jersey, Theodicia Carpenter, who read about Joshee's request for help studying medicine in a missionary newspaper, wrote to Joshee offering to help. Carpenter and Joshee kept up an intimate correspondence, and by the time eighteen-year-old Joshee left India for America in 1883, she thought of Carpenter as her "aunt."[220]

Joshee's quest to study medicine had created something of a stir in her community, and she made a public address at Serampore College before she left, declaring that "there is a growing need for Hindu lady doctors in India, and I volunteer to qualify myself for one."[221] Joshee's ambition was of no less interest in America, where she became something of a celebrity. A devout Hindu, Joshee was the first Indian woman many Americans had ever seen, and certainly the first whose goal was to become a doctor. Joshee was aware of this and cultivated a public image as a kind of ambassador for Hindu women, teaching her missionary friends and fellow medical students about her religion and her culture. She was also one of the first Indians to describe American culture to other Indians.[222]

Joshee entered the WMCP in the fall of 1883, along with a cohort of other international students. Unlike her fellow students, however, Joshee never converted to Protestantism. Similarly, she did not abandon her interest in the Ayurvedic traditions of medicine, and she even wrote her thesis about obstetric practices among Brahmin Hindus.[223] While in school, Joshee contracted tuberculosis and was ill for much of her education. Despite these difficulties, she graduated in 1886, becoming the first Indian woman to receive a medical degree. The completion of her degree was something of a public spectacle; she even received a congratulatory telegram from Queen Victoria.[224] The governor minister of Kolhapur in India asked Joshee to serve as the official "Lady Doctor of

Kolhapur," running the women's ward of the Albert Edward Hospital, and she accepted the position. But on her return to India, her condition worsened, and she was unable to take up a regular medical practice. Joshee died at only twenty-one, having achieved her goal of becoming a physician, even though her larger goal of bringing medical care by and for Hindu women to her home was cut short.[225]

MISSIONARY MEDICINE AND INDIGENOUS COMMUNITIES

American missionaries who took an interest in the possibilities of medical education as a tool for reform were also invested in what they saw as the plight of Indigenous people in North America. Aligning themselves with Progressive Era goals of assimilation for Indigenous people, missionary groups promoted the education of Indigenous doctors who could take Western medicine back to their communities along with Christian religious tenets. The Omaha was one of the Native American tribes subject to such assimilation programs, and as an Omaha physician and reformer, Susan La Flesche Picotte (June 17, 1865–September 18, 1915) would have to navigate a complex and fraught relationship between the Omaha and missionary reformers.

La Flesche was born in a difficult period for the Omaha nation. The dawn of the nineteenth century saw a smallpox epidemic that decimated the Omaha, killing about half of the total population. Fearing for their survival, the Omaha executed a treaty with the United States government, ceding some of their land in exchange for protection.[226] The treaty, and its subsequent misuse by the federal government, kicked off a period of increasing assimilation pressure on the Omaha, as their lands were seized and their livelihood cut off. A decade before her birth in 1865, La Flesche's father was a signatory on another treaty meant to secure a reservation for the Omaha as part of a group who favored assimilation.[227] He invited Protestant missionaries to establish a school on the reservation. Missionaries and other white reformers including anthropologist Alice Fletcher were instrumental in La Flesche's education. La Flesche's early education was a combination of home schooling and formal schooling at the Elizabeth Institute for Young Ladies in New Jersey. At seventeen, La Flesche returned home to teach at a school founded by Quaker missionaries on the reservation. Fletcher secured a place for La Flesche and some of her siblings at the Hampton Institute in Virginia as part of an experimental education program for Indigenous people, from which La Flesche graduated in 1886 after converting to Presbyterianism.[228]

While at the Hampton Institute, La Flesche met the resident physician Martha Waldron, a graduate of the WMCP. Waldron helped La Flesche find funding for medical school at WMCP through the reform organization, the Connecticut Indian Association. As historian Sarah Pripas-Kapit shows, La Flesche's education and contacts with white reformers made her an ideal candidate for continuing her education, and as a future "medical missionary" among the Omaha. White reformers believed that Indigenous professionals educated in Western culture and medicine could return to their people to similarly "uplift" their own.[229] This paternalistic culture of missionary and reform work among Indigenous people was typical of Progressive Era attitudes toward "Indian assimilation."[230]

After graduating at the top of her class from the WMCP in 1889 and completing a yearlong internship, La Flesche took a position as the physician of the Omaha Agency Indian School. Reform groups connected to the school and the Omaha reservation viewed La Flesche as a "medical missionary" who was expected to proselytize while she provided medical care. This mission was not limited to Christian religious teachings and included the promotion of white, middle-class ideas of domesticity and the gendered division of labor. This kind of reform work was intimate, and it involved personally directing the actions of community members to achieve the reform goals of improved hygiene and domestic order. In her own practice, La Flesche sought to "help the women in their housekeeping, teach them a few practical points about cooking and nursing, and especially about cleanliness."[231]

As she became involved in the temperance movement to ban alcohol, her optimism about the benefit of introducing white customs and practices to the Omaha began to wane.[232] After working for a few years at the Indian School, her health was also failing, and she and her husband Henry Picotte moved to Nebraska. She continued to practice in her community, but by the end of the nineteenth century, she came to understand that many of the reform policies she had helped advocate for were actually hurting the Omaha. The frame houses that reformers had encouraged Omaha to build instead of the tipis they claimed were unhygienic caused overcrowding and spread disease. Many of La Flesche's community had become addicted to alcohol, and the corrupt allotment policies of the Omaha Agency had resulted in them losing what little land they had left.[233]

La Flesche, who died on September 18, 1915 after a long period of declining health, has left a complicated legacy as a reformist physician. Even as her faith in the federal government's efforts to assimilate Indigenous people was eroded

SUSAN LA FLESCHE PICOTTE

An Omaha physician and reformer, Susan La Flesche Picotte was educated in a tradition of progressive "missionary medicine" that often conflicted with the needs of her indigenous community.

↓

THE WOMAN'S MEDICAL COLLEGE OF PENNSYLVANIA

A class portrait of medical students from Scalpel, the 1911 yearbook for the Woman's Medical College of Pennsylvania.

by corrupt policies and the decline of the Omaha, she remained allied with religious missionaries and women's reform clubs for the rest of her life. After the federal government declined to fund a hospital La Flesche wanted to found, Presbyterian missionaries stepped in to provide the money.[234] As Pripas-Kapit has observed:

> "[w]hile she had become wary of measures enacted specifically to 'protect' Indians and antagonistic towards the OIA [Office of Indian Affairs], she hoped that general public health and temperance measures supported by progressive women's organizations would help and affect Indians and whites alike."[235]

Women seeking medical education in the nineteenth century navigated an intersection of complex forces, from the professionalization of medicine to colonialism. The dedication required by these women to pursue medical education, often at great cost, was extraordinary, but it was not a simple feat of will. As new technologies made international travel more excessive and missionaries and colonial reformers laid a network of resources and connections

↑ Susan La Flesche Picotte (second row, wearing the brimmed hat) in Decatur, Nebraska c.1910.

across the globe, women were more able than ever before to pursue professional education outside their home countries. While many women were allied to the reformist causes that often supported their educational goals, just as many resisted their paternalism and assimilationist policies and maintained their commitment to providing medical care to their communities. In their struggle to secure respected credentials, women physicians and advocates created an infrastructure of medical education that equaled, and often surpassed, the male medical establishment. The story of women physicians then is not a marginal history of medicine; it is integral to the professionalization of medical practice.

Anandibai Gopalrao Joshee

(March 31, 1865–February 26, 1887)

The woman who was to become India's first woman physician with a medical degree, Anandibai Joshee, was born on March 31, 1865 in Kalyan, Maharashtra, an Indian city-state home to the capital Mumbai. When she was young, Joshee's father, Ganpatrao strayed from orthodox Hindu belief that women should not receive education and encouraged her to go to school. This investment in Joshee's education was continued by her husband, Gopalrao Joshee. They married when Joshee was only twelve, although he was also a troubling partner. Joshee described instances of verbal and physical abuse suffered at his hands in the years they lived together in India.[236]

By fifteen, she was determined to study medicine, a choice likely influenced by the loss of an infant son and surviving a serious bout of illness herself. After years of planning and gaining the support of her community, Joshee set sail from Calcutta on April 7, 1883. Later that year Joshee began training at the Woman's Medical College of Pennsylvania (WMCP, see page 113).

At WMCP, Joshee studied obstetrics and gynecology, hoping to return to India to serve other Indian women. After three years, Joshee graduated with her medical degree and upon graduation, she accepted an offer from the governor minister of Kolhapur in India to serve as "Lady Doctor of Kolhapur" and to run the women's ward at Albert Edward Hospital, a local hospital in Kolhapur.

During her studies, Joshee contracted tuberculosis, and when she returned to India in 1886, her health was in rapid decline. Before she could take up her post at Albert Edward Hospital, she died in February 1887 at the young age of twenty-one. Despite her short life, Joshee's accomplishments were unprecedented for an Indian woman, and her achievements were enough to open the door for other Indian women to quickly follow.

Section IV

The Twentieth Century, Pre-World War II

CHAPTER 11

"Powerful levers that move worlds!"

THE WOMAN COMPUTER

In 1899, astronomer Dorothea Klumpke (August 9, 1861–Octobere 5, 1942) of the Paris Observatory traveled to London for the International Congress of Women, a gathering of suffrage groups from around the world. From June 26 to July 7, women from various professions spoke on topics ranging from women's education to politics to women in industry. In the Small Hall in the Westminster Town Hall on June 29, the Science Section of the Congress held court, with Klumpke representing women in astronomy. In her speech, "Women's Work in Astronomy," Klumpke identified three distinct historical periods of women in astronomy: the first in antiquity, the second in the Enlightenment and Renaissance, and the third in Klumpke's present day at the turn of the twentieth century. "Day by day, women bending over the micrometers examine the photographic skies and measure the star-positions to form the catalogue which is to be a legacy of our century," she said. "At a few of the national observatories woman has observed planets, comets, and shooting-stars [...] among the toilers of the various branches mentioned above are found the qualities requisite for producing lasting results— concentration and enthusiasm, powerful levers that move worlds!"[237]

What Klumpke described in this third period of women in astronomy is the era of the "lady computer." From 1880 to about 1930, hundreds of lady computers toiled away in observatories all around the world: "Paris, the Cape of Good Hope, Helsingfors. Toulouse, Potsdam, Greenwich, Oxford," Klumpke listed them. With the unprecedented number of women in astronomy during this period, Klumpke said, "a new element—*equality*—appears." From Klumpke's view, this new period likely did resemble equality: she had

← Women computers at the Harvard College Observatory in 1890.

risen in the ranks of professional astronomy to be Director of the Bureau of Measurements at the Paris Observatory with five computers under her supervision. But most women who entered the field of astronomy as a computer rarely had the opportunity to become more.

The job of the computer was to perform calculations for scientific research. The solitary computer has a long history, starting with Maria Cunitz's simplification and correction of Kepler's *Rudolphine Tables* in her magnum opus, *Urania Propitia*, who was an early example of one woman performing this kind of work (see pages 37–38).[238] When computers worked together in organized groups, however, they had the most influence as integral parts of more modern projects of "big science."[239] Klumpke's reference to "the catalogue which is to be a legacy of our century" was one such project, the Astrographic Catalogue (AC). The AC was initiated in 1887 at Klumpke's home observatory in Paris, and it enlisted the work of twenty observatories worldwide to photograph, measure, and catalogue the entire night sky. Such an ambitious project required untold hours of labor to complete, much of it involving meticulous, repetitive, and tedious calculation. A vast workforce of computers was required for such an undertaking.

↑ The Paris Observatory in Meudon, in Hauts-de-Seine, France.

One of the earliest instances of computers working together as a group comes from the eighteenth century, when Nicole-Reine Lepaute, Jérôme Lalande, and Alexis-Claude Clairaut set to work charting Halley's Comet's path around the sun and predicting its return (see page 39).[240] The way they divided up the labor by type of calculation would provide a model for larger groups later on: Clairaut worked specifically on the orbit of the comet, Lepaute and Lalande handled the three-body problem of Saturn and Jupiter's gravitational exertion on the comet, and Clairaut would check their final computations for errors. At the end, the individual pieces were compiled into their final calculation of the date of the comet's perihelion.[241]

Beginning around the 1880s and continuing through the early twentieth century—until electronic computers and calculation devices replaced them—women computers became a common fixture in observatories around the world. Three main coinciding factors created this pool of women workers. First, there was a surge in college-educated women who were graduating from newly created women's colleges. Second, discriminatory hiring practices kept women out of professional positions in science in universities and government jobs, making work as computers one of a very limited number of employment options. Finally, changes in the structure of science created "big science" with larger budgets and more support staff, creating positions that could be filled by women graduates looking for work.[242] Working as a computer gave women the opportunity to do meaningful work that harnessed their hard-won college education, but the work was not necessarily prestigious, nor were the conditions particularly agreeable.

COMPUTERS OF THE HARVARD OBSERVATORY

Computers were one of the lowest-ranking positions at the observatory, which meant there was dismal pay and little to no room for advancement. Their work was often segregated from that of men, who were more engaged with the work of observation and leadership roles at the observatory. Hiring women for these positions had little to do with a progressive belief in women's equality and more to do with tapping into a new employee pool that observatories could acceptably pay poorly to perform huge numbers of computations. Confined to these duties, women would not threaten the positions of the men already in the profession. When Edward Charles Pickering, Director of the Harvard Observatory, filled his observatory with women computers, he had that very thought in mind. In the 1898 *Annual Report of the Harvard Observatory*, he wrote

"To attain the greatest efficiency, a skillful observer should never be obliged to spend time on what could be done equally well by an assistant at a much lower salary."[243]

Harvard hired its first woman computer, Anna Winlock in 1875. The computer program expanded under Pickering when he hired Williamina P. Fleming, his housekeeper, six years later. The program grew again in 1883 when Pickering began the Henry Draper Memorial, a large-scale project with the aim of photographing the stars and determining their spectral classification. Funded by Anna Draper, Henry Draper's widow, Fleming and Pickering hired twenty women between 1885 and 1900, including the Drapers' niece Antonia Maury, Annie Jump Cannon, Henrietta Swan Leavitt, Margaret Harwood and many more in the twentieth century.

Secluded in their own room in the observatory, the computers would spend each day poring over photographic glass plates of the night sky, calculating the positions of stars, measuring their distance, and determining their brightness. This work was repetitive, much the same every hour of every day, but it also required accuracy. Women had to be disciplined and pay attention to detail with every new plate placed before them. Despite the repetitiveness and low-status of their position, many of these women broke new ground in this niche. Fleming, whose original duties included copying and basic computation, discovered ten novae, more than three hundred variable stars, and fifty-nine nebulae. She developed the Observatory's first classification system of stellar spectra. Maury improved Fleming's classification and developed a more expansive one of her own. Cannon classified more stars than any single person, an estimated 350,000 over her decades-long career, and discovered 300 variable stars and five novae. She united the two previous classification systems into the Harvard Classification Scheme, still used today. And Leavitt, the discoverer of over 1,777 variable stars, was the first person to explain the relationship between the variation in the magnitude of a star's brightness and its period of variation, the key for measuring distance through space.

Compared to other observatory directors, Pickering was rather supportive of his computers and their work: he co-published papers with them and publicly valued their accomplishments. But his progressiveness had its limits. From the time of Anna Winlock's hiring in 1875 to 1906, women made only twenty-five cents per hour. This was particularly upsetting to Fleming, who was a single mother. In a journal entry in March of 1900, Fleming vented her frustration, writing, "[Pickering] seems to think that no work is too much too hard for me, no matter what the responsibility or how long the hours...Sometimes I feel tempted to give

↑

WILLIAMINA P. FLEMING

Williamina Fleming was one of the first Harvard computers
and was appointed Curator of Astronomical Photographs. She
discovered ten novae, more than three hundred variable stars,
and fifty-nine nebulae, including the Horsehead Nebula.

↑

ANNIE JUMP CANNON &
HENRIETTA SWAN LEAVITT

Annie Jump Cannon (left) developed a system for classifying stellar
spectra and classified an estimated 350,000 stars in her career.
Henrietta Swan Leavitt specialized in variable stars and developed the
period-luminosity relationship, also called Leavitt's law, which is an
essential tool for measuring distance through space.

up and let him try some one [sic] else, or some of the men to do my work, in order to have him find out what he is getting for $1,500 a year from me compared with $2,500 from some other [male] assistants...But I suppose a woman has no claim to such comforts. And this is considered an enlightened age!"[244]

Despite her anger, Fleming stayed at the observatory until her death in 1911. In a biographical profile of Fleming in *The New England Magazine* published the following year, the writer summed up the role of women's work in astronomy in what was no doubt intended to be glorifying terms:

> *"It seems that that very genius for tactful execution, for patient attention to detail, for swift comprehension, which has re-enforced [women's] intelligence in astronomy and lifted them into prominence, has been only a larger expression of the power that made them able housekeepers and home-makers. For women, in the world of science, wherever notably successful, have not come there as rivals of men, but rather have supplemented and extended, often suggested and planned the work of men, thus fulfilling the principle of their mission as helpmeets."*

Even in light of all their groundbreaking work, women computers were still defined by their relationship to men.

Women's work as computers did not just encompass the rote calculating, measuring, and cataloguing that the workers did every day but also described their secondary role in relation to their male superiors—a role that Maury would not endure. She could not reconcile her desire to take up more theoretical work against Pickering's wishes to keep her engaged in "women's work," so she left the computer group in 1892.[245] After she left, she wrote to Pickering demanding to be made full author for her classification system, even though Fleming had only received an acknowledgment for her own. "I worked out the theory at the cost of much thought and elaborate comparison," she wrote, "I should have full credit for my theory."[246]

THE RISE AND FALL OF THE GREENWICH OBSERVATORY COMPUTERS

As an experiment in economizing the observatory's work, the women's computer program at Harvard was a success, and it set a precedent for other observatories around the world to create their own programs. Like Harvard, these observatories created new jobs for women in astronomy, but they also reproduced Harvard's discriminatory hierarchy and pay practices. In 1889, the Greenwich Observatory in Britain, under the leadership of the Astronomer Royal William Christie, formed a "lady computer" program when the observatory increased its budget for computers by forty percent.[247]

These new computers were the first women ever hired at Greenwich, a fact that reflects how few opportunities there were for women in astronomy at the turn of the twentieth century. Even women who graduated from women's colleges and took university examinations found themselves with few to zero job options. Typically, students at women's colleges in Britain did not receive astronomy-specific education. It was integrated into mathematics and physics programs, and while this gave them solid skills for computing, this curriculum limited their chances of finding work in astronomical observation.[248] Professional astronomy also prized hands-on experience with observation, but women rarely had access to telescopes and other instruments, especially outside universities.

Greenwich hired its first three computers, Edith Mary Rix, Harriet Furniss, and Alice Everett, in 1890 and Annie Scott Dill Russell in 1891. In Greenwich's staff structure, these women were considered "supernumerary computers." This was the lowest-ranking position at the observatory from which there was no hope to advance because the observatory operated under the Civil Service, which prohibited women from permanent service.[249] Previously, this role had been filled by teenage boys, thirteen to fourteen years old, who were expected to leave for other professions after several years. Each of these women had college degrees, and both Everett and Russell obtained honors in Cambridge's famously difficult Tripos mathematics exam. But the women computers served the same function as the teen boys and were all paid the same, four pounds a month, save for Everett who made six. When Russell learned she would only make four pounds a month, she could hardly believe that her education at Girton College and success on the Tripos at Cambridge had earned her so little. She wrote to Herbert Hall Turner, the observatory secretary, saying, the amount "is so small that I could scarcely live on it...that I have taken the Mathematical Tripos at Cambridge makes no difference?"[250] It did not. Furniss left the observatory just one year into her term, and Rix quickly followed the next year. Although Greenwich tried to replace them, they found no one willing to accept such low pay and rank, and just two years into the program, only Everett and Russell remained.[251]

Everett and Russell worked the same hours as everyone else in the observatory: 9:00 am to 1:00 pm every weekday; 2:00 pm to 4:30 pm three afternoons a week; and three nights a week for two to four hours.[252] Their days were full as they bounced between departments working on data recording, astrophysical research, meteorology, meridian transit work, and solar photography. Unlike the Harvard computers, Everett and Russell were not segregated from the men and used observational instruments in some of their work. In 1892, Everett began work on the international Astrographic Catalogue (AC), developing photographic plates and measuring them for the project, becoming the first person to make a series of AC measurements at Greenwich. Among her other duties, Russell was responsible for the daily measuring of sunspot photographs under Walter Maunder, First Assistant for Photographic and Spectroscopic Observations.

In 1891, Everett and Russell attempted to join the Royal Astronomical Society (RAS) as fellows, with support from Maunder. But aside from a select few, women were still excluded from the RAS. In January of 1892, the RAS officially rejected their election. As a consolation prize, the RAS permitted the women to attend their meetings, but Everett sought opportunity elsewhere.

↑ Edward Pickering, director of the Harvard College Observatory, with the computers in 1911.
Top row (left to right): Margaret Harwoody, Mollie O'Reilly, Edward Pickering,
Edith Gill, Annie Jump Cannon, Evelyn Leland, Florence Cushman, Marion Whyte.
Bottom row (right to left): Grace Brooks, Arville Walker, Johanna Mackie, in front
of Pickering Alta Carpenter, Mabel Gill, and Ida Woods.

She first applied to work in the stellar photography program at Dunsink Observatory in Dublin, which passed her up for a less-experienced man ten years younger, and second to the Astrophysical Observatory in Potsdam, Germany. Potsdam accepted her, and in 1895, she became the first employed woman astronomer in Germany. That same year Russell was forced to resign when she married Maunder because the Civil Servicedid not allow married women to work, a ban that remained in place until 1946. With Everett and Russell gone and no other women willing to fill these positions, the short lived "lady computer" program at Greenwich ended.

AUSTRALIA'S COMPUTERS AND THE ASTRONOMICAL CATALOGUE

Along with Greenwich, four observatories in the Australian cities of Perth, Melbourne, Sydney, and Adelaide were invited to join the AC. Like Greenwich, they hired women computers. Together, the four observatories covered eighteen percent of the sky for the entire AC, so they needed all the hired help they could get. From 1891 to 1963, the four observatories employed sixty-one women, who were variously called "astrographic assistant," "star measurer," "clerk," "junior computer," or "astrographic computer."[253] The stories of their careers have been difficult to bring to light because they were largely unnamed and unacknowledged. Local media reports about the AC usually referred to the women computers merely as "girls" or "ladies," and only three scientific papers that resulted from the Australian contributions to the AC acknowledged individual women.[254] In the annual reports of the four observatories, they were just called "measurers."[255] Computers had become an anonymous workforce, confined to an under-class by low pay and no advancement.

Mary Emma Greayer was one such computer whose life has been recovered. Employed at the Adelaide Observatory in 1890, Greayer worked on nighttime observations and the reduction of hundreds of stars. She was the main observer of the individual zenith stars, and between 1894 and 1898, she personally observed more than one third of the positional stars for the Melbourne zone in the AC.[256] In 1893, Greayer became one of the first women to join the Astronomical Society of South Australia, but like many other women, left the observatory when she married in 1898. Another, Charlotte Emily Fforde Peel, worked at the Melbourne Observatory from 1898 to 1918, measuring stars, reducing star coordinates, and checking errors of other workers.[257] In 1900, Peele was made a permanent staff member at Melbourne, becoming the first woman to be permanently employed in astronomy in Australia.

To prevent the risk of the women computers distracting male employees from their work, they spent most of their working hours segregated. Typical hours for women computers in Australian observatories ran 9:00am to 5:00pm on weekdays and 9:00am to noon on Saturdays, with some exceptions for people like Greayer who observed nights. For a more than forty-hour work week, they were paid forty pounds per year, nearly half of what the men were paid. This pay disparity was even legalized in 1902 by the Commonwealth Government Act, which legislated that women be paid no more than fifty-four percent of a man's salary. In 1913, four computers at the Perth Observatory, Prudence Williams, Minnie Harvey, Ethel Allen, and Ida Tothill, advocated for higher pay. Surprisingly, they were granted their request for higher pay as well as a one year extension on their original three year contracts.[258]

When Dorothea Klumpke spoke on women's work in astronomy at the International Congress of Women she was right in one respect: women were, indeed, hard at work in observatories all over the world and their work was the "powerful levers that move worlds." But true equality—equal pay, equal rank, equal respect—had not yet arrived. The work of women computers was tedious, seemingly endless labor, and for all its complexity, professional male astronomers often devalued it as grunt work not worth their own time.

But this was the work on which the early big science projects in astronomy were built—theirs was the force that kept the gears turning and the telescopes sweeping. Before the Harvard program, variable stars did not inspire much interest in the astronomical community, but the use of dry-plate photography and the work of women computers initiated a shift in the discipline. By 1959, women had discovered more than seventy-five percent of the 14,708 variable stars known to date.[259] This particular area of astronomy became defined as women's work simply because it was women who did it. Even in the face of such blatant marginalization, it was in these subordinate positions that innovative and extraordinary work happened.

CHAPTER 12

The Home as Laboratory

At the turn of the twentieth century, enthusiasm for the social and economic benefits of science and technology reached a fever pitch. All over the world, but particularly in the United States, the so-called Progressive Era was characterized in part by the widespread embrace of scientific and technological principles, from manufacturing to missionary work. But even as the Progressives extolled the potential of innovation and scientific thinking to change society for the better, women were still largely barred from participating in professional scientific and technical work. Instead, women created space to participate in science within their own "separate sphere" of domestic life, incorporating science into the way they cleaned their homes, parented their children, and prepared food.

Some women went a step farther and conducted science in their own home-based experiment stations and laboratories. By incorporating scientific ideas and practices into their everyday lives, many women participated in science on their own terms, creating space for themselves and others at the margins of science. When we look at the scientifically-managed home and the home laboratory as places where scientific knowledge was produced and valued, we better understand the ways that women at the margins of the scientific enterprise nevertheless fully participated in the progressive wave of innovation and reform.

DOMESTIC ENGINEERING

In 1913, the US edition of the women's magazine *Good Housekeeping* published an article by Harriet Gillespie, a New Jersey woman who claimed to have

=

lived a full year without any servants and encouraged others to consider doing the same.[260] In her home "experiment station," Gillespie tested out both new technologies for managing the endless labor of housekeeping and new scientific methods of reducing unnecessary motion. Gillespie enthusiastically advocated for women to take up these new approaches to standardizing housework to reap the benefits of a reduced household budget by letting servants go. After all, it was apparent to many middle-class women that the young woman servant "was fast eliminating herself," Gillespie argued. "She is going into the shop, the factory or other industry, where she can have her Sundays off, her regular hours of work, her time for recreation and her feeling of independence."[261]

Gillespie described how she was able to take on her own housework by utilizing new technologies, such as "pneumatic cleaners for the floors, walls and furniture" and a washing machine, which, despite an initial high cost, "returned eighty percent on the investment in the first year." And if the reader needed more convincing of the benefits of becoming a "domestic engineer," she "advise[d] every woman to read what those experts, Harrington Emerson, Frederick Taylor and Frank Gilbreth, have to say on the subject of efficiency, and then translate those principles to their homes. In that way each woman can start herself toward an appreciation of the value of domestic engineering."

Domestic engineering, domestic science, and home economics are all part of a larger history of changing social and economic structures at the turn of the twentieth century, changes that brought new science and technology right into the "separate sphere" of the middle-class home. The scientific management of labor and production was revolutionizing industry. Engineers including Frederick Winslow Taylor, who Gillespie singled out, developed and applied scientific theories of efficiency to fields like manufacturing, among others, with the goal of increasing economic output. Taylor believed that inefficient work practices could be curtailed by removing the idiosyncrasies individual workers brought to their tasks. This could be accomplished in part by assigning planning work to managers, who would delegate the tasks to employees, and standardizing the way those tasks were executed across all workers. This strict division between management and labor, combined with scientifically tested methods for completing work tasks, would result in increased efficiency and profit for companies.

Taylor and other scientific management advocates believed there was one ideal way to perform any given work task, and by studying workers and measuring the time it took them to complete tasks and the quality of their work, scientific managers could derive that one best way and make all workers use the method. Taylor's ideas were not well received at first in a labor culture that valued skill and the individual expertise of workers, but as the increased production associated with scientific management became more widely known, companies and foreign governments showed more interest. "Taylorism" became popular, however, outside of industry, as part of the larger progressive social

By incorporating scientific ideas and practices into their everyday lives, many women participated in science on their own terms, creating space themselves and others at the margins of science.

and political movements of the period, in which the application of science and technology to modern life promised efficiency, prosperity, and social reform.

As historian Elisa Miller has shown, the introduction of science and technology to the management of the domestic sphere was something of a second wave in the history of home economics. An earlier vision of the rational management of the home came from religious missionary reform, which emphasized the moral and spiritual benefits of a well-kept home and its role in preserving traditional family structures. With the introduction of scientific management to the social and cultural landscape of the late nineteenth century and amid increasing opportunities for women to attend college, home economics became a distinct academic discipline that embraced scientific and technical progress in the realm of homemaking.[262] Students in home economics learned as much biology and chemistry as they did cooking to better embrace innovations in sanitary theory in keeping their homes. And, as Gillespie suggested, many of the same scientific managers and engineers who consulted with companies to improve their efficiency also advocated for the application of those principles to the home. As part of the progressive drive to improve society at every level, people like Gillespie were enthusiastic about the possibilities of science to improve the management of the home and the fundamental structures of domestic life.

As scientific management was becoming more and more popular in industry, many women writers were advocating for the application of those principles to the domestic sphere.[263] In 1915, a woman named Mary Pattison published *The Principles of Domestic Engineering*, a report on the results of studies conducted at the same home-experiment station that Gillespie ran in New Jersey. In the preface, Pattison wrote that the aim of the book "...was to meet what has been generally termed the 'Servant Problem.'" Like Gillespie's *Good Housekeeping* article, *Principles* advocated for the use of new household technologies listed at the end of the book. Pattison covered the practicalities of the domestically-engineered home, such as making a budget and a system of inventorying household items, and she laid out the social reformist aims of eliminating servants and elevating housework from its status as menial labor. Although advocates for domestic engineering, such as Pattison, likened the elimination of servants to the abolition of slavery, many also believed that servants were incapable of learning the scientific principles that underlay domestic engineering; another reason it was better for housewives to do things themselves.[264]

This sentiment is echoed in the same issue of *Good Housekeeping* in which Gillespie's article appears. Letters to the editor on the subject of adopting new technologies in the home make it clear that domestic engineering, for all its pretensions to class solidarity, reflected the dominant class and racial stratification of Progressive society. One letter writer, for instance, wrote to offer

Women made space for science in their homes, and as a result, were sometimes able to enter into the male-dominated scientific establishment on their own terms.

advice on how to integrate new technology in the home when one's servants were "colored." "Most housekeepers," wrote a reader from Alabama, "consider this class of labor unsatisfactory and stupid in the use of labor-saving household devices and so some of them are."[265] While domestic engineering and home economics taught women about science and its application to the home, this type of opportunity was still one of relative privilege. For women who were too poor to attend college or even purchase books about domestic engineering, the drudgery of housework continued unabated by the shiny gloss of scientific management. And white immigrant women and women of color who worked as domestic servants often did not have their own homes to keep, much less manage scientifically.

THE GILBRETH SYSTEM

Frank Gilbreth, another expert Gillespie recommended to aspiring domestic engineers, and his wife Lillian (May 24, 1878–January 2, 1972) published some of the most important studies in the emerging field of scientific management in the early twentieth century. Born Lillian Moller in Oakland, California, in 1878, Lillian came from a large, wealthy family. Her father owned a successful business, and her parents employed several domestic servants in the family home.[266] She was educated at University of California, Berkeley, where she earned a master's degree in literature. Later when her family moved to Providence, Rhode Island, she enrolled in the doctoral program in Applied Psychology, receiving her PhD in 1915. Her plans to become a university dean were altered after her marriage to construction contractor Frank Gilbreth in 1904, and the couple began planning both a large family and a business.[267]

Together, Lillian and Frank began writing books that expanded on some of the methods for efficient construction that Frank had developed in his own construction work. The Gilbreths built upon Taylor's "stopwatch" studies, which broke a work task into discrete parts and measured the time it took to complete each part. They developed a system of "motion studies" using film technology to trace the motion of workers as they completed tasks.[268] Beginning in 1912, the Gilbreths moved even further away from the Taylorism that had ignited the scientific management movement, with Lillian in particular using her training to address the psychological dimensions of management.[269] Their new "Gilbreth System" involved motion studies to eliminate unnecessary movement as opposed to simply exhorting workers to perform tasks faster, and it attended to worker dissatisfaction that had resulted from Taylorism's tendency to dehumanize workers and depersonalized their work.[270]

In 1924, after more than a decade of successful business partnership and co-authorship of numerous books, lectures, courses, and articles, the Gilbreths' marriage ended in tragedy. Frank died of a heart attack at fifty-five, leaving

↑

LILLIAN MOLLER GILBRETH

Psychologist and industrial engineer Lillian Moller Gilbreth developed
new methods of time and motion management for industry alongside
her husband Frank GIlbreth. After his death, Lillian re-started her
career as an industrial designer, pioneering new labor-saving kitchen
designs and methods for household management.

Lillian and her eleven children with an uncertain income. In the years following Frank's death, Lillian found that the consulting business they had built together could not survive without a man as its public face. Clients declined to renew their contracts or cancelled them outright, and new clients balked at hiring a woman industrial consultant.[271]

According to historian Laurel Graham, Lillian recovered her career and improved her prospects in the 1920s by reinventing herself as an industrial designer. Lillian began taking commissions to design efficient kitchens for utility companies invested in selling gas and electric appliances.[272] Gilbreth used principles from motion studies to arrange appliances, cabinets, and work surfaces in the most efficient layouts, and her clients marketed the kitchens with pamphlets describing her expertise and how the new designs would make work in the kitchen easier and more enjoyable.[273] Despite her focus on design coming late in her career, these projects were among Lillian's most profitable endeavors, and the large commercial reach of the model kitchen designs ensured that she became an important figure in the history of scientific management in the home. As Graham notes, Lillian's image as a famous homemaker and kitchen designer was more a product of the marketing of the kitchens, many of which featured images of her with her many children, than a reflection of her actual home life. Often far too busy to do housework herself, she employed domestic servants and a system for chores performed by her children to ensure she had time for her career. But in marketing images, she was portrayed as doing all this work herself, something that consumers were encouraged to aspire to in buying appliances and adopting her methods.

Domestic engineering, home economics, and other programs for the scientific management of the home were one way in which women participated in science despite social norms that sought to limit them to the domestic sphere. These fields are an example of the way that enthusiasm for scientific and technical progress infused all aspects of life at the turn of the century, and they demonstrate the many ways that women sought out and used the tools provided by science to better their own lives. But there were also women who did what we might more readily recognize as scientific work outside of the laboratories and research institutions from which they were typically banned. Women made space for science in their homes, and as a result, were sometimes able to enter into the male-dominated scientific establishment on their own terms. At the same time domestic engineers were reshaping home life with scientific principles, women like Abbie Lathrop were reshaping science from within their domestic spaces, and in Lathrop's case, her barn.

ABBIE LATHROP'S DOMESTIC LABORATORY

Abbie Lathrop was born in 1868 in Illinois. After very little formal education and a brief stint as a schoolteacher cut short by illness, she moved to Massachusetts to start a doomed poultry farming business. Where she failed at breeding chickens, Lathrop would succeed—spectacularly so—in breeding mice and rats. At first a "fancier," that is someone who breeds pet and show animals for specific appearance or behavior, Lathrop's mouse-breeding operation became a fully-fledged research operation after she began selling mice to scientists in need of laboratory subjects.[274] Mice were good candidates for laboratory research because of their short lifespans, and those obtained from fanciers could be custom-ordered with specific genetic traits that made it simpler for scientists to control variables in their experiments.[275]

The evolution of Lathrop's mouse-breeding business into a scientific enterprise began in 1908 when she noticed that some of her mice had developed

↑ A woman in her kitchen in Texas in 1939. The design of efficient kitchens was an important way that science and technology entered the home in the twentieth century.

skin lesions. Well-connected to scientific and medical researchers who purchased mice from her, she enlisted their help to determine the cause of the lesions. Leo Loeb, a pathologist at the University of Pennsylvania who used Lathrop's mice in his work, was able to confirm that the lesions were cancer. This incident sparked a long-running collaborative research project, in which Loeb and Lathrop co-authored a number of papers about the tumors Lathrop had detected in her mice.[276]

The experiments that Lathrop and Loeb designed were performed by Lathrop at her home, transforming her farm into a laboratory.[277] Lathrop's unusual entry into the scientific profession, both as a woman and someone with very little formal education, made her an object of much fascination. The *Brooklyn Eagle* profiled her in 1909, noting that "[s]he is one of the few women who have an instinctive liking for these little creatures."[278] Similarly the *Los Angeles Times* reported Lathrop saying she *was* afraid of rats and mice, but overcame her fear "when she found there was money in them."[279]

For white, upper- and middle-class women in the Progressive Era, opportunities to participate in science were limited, but certainly not non-existent. Women like Harriet Gillespie and Mary Pattison created whole systems of scientifically-minded homemaking that embraced new theories of sanitation and new technologies that automated the drudgery of things like laundry. Their work responded to the changing economic conditions as more and more domestic servants turned to factory work for better wages. Lillian Gilbreth, who began her career working as a consultant in industry, leveraged the expectations of domestic life and motherhood to maintain her career. And Abbie Lathrop made significant contributions to the study of cancer from her farmhouse in New England, even though she had never attended college. These women exemplify some of the ways in which women have always made space to practice science outside the mainstream establishment, using gendered social and cultural strictures about the "place" of women as resources to forge their own paths.

CHAPTER 13

Women's Reproductive Freedom and the Eugenics Movement

NEW RADICAL WOMEN

The turn of the twentieth century marked a moment of rebellion for women in the West, particularly in Europe and the United States. The first wave of feminism was underway, bringing with it militant activism in reproductive rights, the right to vote, and labor reform. Political movements for socialism and anarchism attracted women who were discontented with their lot in a strictly patriarchal and increasingly industrialized society. For some, the cautious middle-class approach to reform through religious and voluntary organization was too politically conservative and the progress promised by the "Progressives" was laden with backward-looking bourgeois values. The new radical women activists pushed for immense change and were often met with intense opposition.

Reproductive freedom and family planning, what would be called the "birth control" movement, was among the most radical propositions of feminism. Women doctors and radical activists led the struggle for birth control in Britain and the United States, and to bolster their cause, many of them harnessed a transatlantic enthusiasm for scientific solutions to social problems. One branch of the birth control movement, however, eventually coalesced into the eugenics movement, a branch of genetics that sought to improve the human condition through controlled breeding. The story of the birth control movement is the story of the complex role of science in the social and political life of the early twentieth century.

← Suffragette and birth control activist Kitty Marion selling the *Birth Control Review* in New York City in 1915.

Well ahead of the curve in Britain and the United States, the world's first birth-control clinic was opened by Dutchwoman Aletta Jacobs (February 9, 1854–August 10, 1929). She was the first woman in The Netherlands to attend a university and receive a medical degree. Jacobs was a physician, suffragist, birth-control advocate, and peace activist. Despite resistance from the medical community, Jacobs opened her clinic, and in serving women there, she combined her advocacy for women's healthcare and for women's rights in labor, seeing firsthand the toll long work days in deleterious conditions took on women's bodies. Beyond medicine, Jacobs co-founded the Woman Suffrage Alliance and helped establish the Women's International League for Peace and Freedom.

Dutch birth control advocates, like those in Britain and the United States, campaigned for contraceptives for a variety of reasons. Jacobs affiliated with of the World League for Sexual Reform (WLSR), a international group focused on progressive sex reform. According to historian Henny Brandhorst, the Dutch section of the WLSR was small, but its membership believed in sex reform for social, economic, and sometimes eugenic reasons. The debates about the eugenic uses of birth control in the early twentieth century would become much more visible in Britain and the United States, where advocates and reformers rose to public prominence campaigning for contraceptives and family planning.

In Britain, the most famous birth control advocate was Marie Stopes (October 15, 1880–October 2, 1958). Born Marie Charlotte Carmichael in 1880, Stopes grew up in Edinburgh with parents Henry Stopes, a brewer and amateur scientist, and Charlotte Carmichael Stopes, a feminist activist and Shakespeare scholar. Like her father, Marie was interested in science, and she earned several degrees, including a BSc with double honors, in botany and paleobotany at University College London. In 1905, at the age of twenty-four, Stopes became the youngest doctor of science in Britain. After a distinguished education at English and German universities, Stopes took a post as a lecturer in Botany at the University of Manchester and began a remarkable career as a scientist. Although Stopes achieved much as a botanist, including two years of research and expeditions in Japan, writing numerous important papers, and election to the Linnean Society, she has become far more well known for her influence in the birth control movement in Britain.[280]

In 1915, Stopes, by then a distinguished scientist and committed suffragist, met the American birth control activist Margaret Sanger (see page 153). Sanger had fled to Britain to escape charges for violating the Comstock Laws, a set of federal acts prohibiting the circulation of "obscene" materials, including the distribution of information on contraceptives. Stopes had been working on a research project to determine whether there was a measurable cycle of sexual arousal for women by making detailed records of her own feelings. The project culminated with the publication of *Married Love: A New Contribution to the Solution of Sex Difficulties* in 1918, followed that same year by *Wise Parenthood: A Book for Married People*.[281] Both books secured Stopes's reputation as a trusted advisor on matters of sex, marriage, and family. As Stopes's interests shifted from botany to sexuality and family life, she turned her attention to the question of contraceptives and the social and political dimensions of birth control.

THE EUGENICS MOVEMENT

Among those in Britain who shared Stopes's interest in the social dimensions of reproduction were feminist activists and suffragists, members of the government and clergy, scientists and physicians, and a new group of scientific and reform-minded campaigners called eugenicists. At the turn of the century, it was widely believed that Europe was suffering from a precipitous decline in the mental and physical constitution of the "race," that is white people. Eugenics promised simple solutions to complex social problems, but its proponents

The story of the birth control movement is the story of the complex role of science in the social and political life of the early twentieth century.

were fueled by racist fears about increasing immigration, beliefs about the profligate sexuality of the poor, and the weakening effects of the stresses of industrialization and modernization on the middle class.

Eugenics remains a difficult movement to describe. Its adherents often held complex, sometimes contradictory beliefs, and eugenics itself is more of a constellation of ideas, a combination of science and popular social beliefs that were touted as an objective science grounded in the hard truths of genetics and

THE
WOMAN
REBEL

NO GODS NO MASTERS

VOL. I.   MARCH 1914   NO. 1.

THE AIM

This paper will not be the champion of any "ism."

All rebel women are invited to contribute to its columns.

The majority of papers usually adjust themselves to the ideas of their readers but the WOMAN REBEL will obstinately refuse to be adjusted.

The aim of this paper will be to stimulate working women to think for themselves and to build up a conscious fighting character.

An early feature will be a series of articles written by the editor for girls from fourteen to eighteen years of age. In this present chaos of sex atmosphere it is difficult for the girl of this uncertain age to know just what to do or really what constitutes clean living without prudishness. All this slushy talk about white slavery, the man pointed and described as a hideous vulture pouncing down upon the young, pure and innocent girl, dragging her through the medium of grape juice and lemonade and then dragging her off to his foul den for other men equally as vicious to feed and fatten on her enforced slavery — surely this picture is enough to sicken and disgust every thinking woman and man, who has lived even a few years past the adolescent age. Could any more repulsive and foul conception of sex be given to adolescent girls as a preparation for life than this picture that is being perpetuated by the stupidly ignorant in the name of "sex education"?

If it were possible to get the truth from girls who work in prostitution to-day, I believe most of them would tell you that the first sex experience was with a sweetheart or through the desire for a sweetheart or something impelling within themselves, the nature of which they knew not, neither could they control. Society does not forgive this act when it is based upon the natural impulses and feelings of a young girl. It prefers the other story of the grape juice procurer which makes it easy to shift the blame from its own shoulders, to cast the stone and to evade the unpleasant facts that it alone is responsible for. It sheds sympathetic tears over white slavery, holds the often mythical procurer up as a target, while in reality it is supported by the misery it engenders.

It, as reported, there are approximately 35,000 women working as prostitutes in New York City alone, is it not sane to conclude that some force, some living, powerful, social force is at play to compel these women to work at a trade which involves police persecution, social ostracism and the constant danger of exposure to venereal diseases. From my own knowledge of adolescent girls and from sincere expressions of women working as prostitutes inspired by mutual understanding and confidence I claim that the first sexual act of these so-called wayward girls is partly given, partly desired yet reluctantly so because of the fear of the consequences together with the dread of lost respect of the man. These fears interfere with mutuality of expression —the man becomes conscious of the responsibility of the act and often refuses to see her again, sometimes leaving the town and usually denouncing her as having been with "other fellows." His sole aim is to throw off responsibility. The same uncertainty in these emotions is experienced by girls in marriage in as great a proportion as in the unmarried. After the first experience the life of a girl varies. All these girls do not necessarily go into prostitution. They have had an experience which has not "ruined" them, but rather given them a larger vision of life, stronger feelings and a broader understanding of human nature. The adolescent girl does not understand herself. She is full of contradictions, whims, emotions. Her hot emotional nature longs for caresses, to touch, to kiss. She is often so well satisfied to hold hands or to go arm in arm with a girl as in the companionship of a boy.

It is these and kindred facts upon which the WOMAN REBEL will dwell from time to time and from which it is hoped the young girl will derive some knowledge of her nature, and conduct her life upon such knowledge.

It will also be the aim of the WOMAN REBEL to advocate the prevention of conception and to impart such knowledge in the columns of this paper.

Other subjects, including the slavery through motherhood; through things, the home, public opinion and so forth, will be dealt with.

It is also the aim of this paper to circulate among those women who work in prostitution; to voice their wrongs; to expose the police persecution which hovers over them and to give free expression to their thoughts, hopes and opinions.

And at all times the WOMAN REBEL will strenuously advocate economic emancipation.

THE NEW FEMINISTS

That apologetic tone of the new American feminists which plainly says "Really, Madam Public Opinion, we are all quite harmless and perfectly respectable" was the keynote of the first and second mass meetings held at Cooper Union on the 17th and 20th of February last.

The ideas advanced were very old and time-worn even to the ordinary church-going woman who reads the magazines and comes in contact with current thought. The "right to work," the "right to ignore fashions," the "right to keep her own name," the "right to organize," the "right of the mother to work"; all these so-called rights fail to arouse enthusiasm because to-day they are all recognized by society and there exist neither laws nor strong opposition to any of them.

It is evident they represent a middle class woman's movement; an echo, but a very weak echo, of the English constitutional suffragists. Consideration of the working woman's freedom was ignored. The problems which affect the

---

←

THE WOMEN REBEL

The first issue of Margaret Sanger's birth control publication *The Woman Rebel.*

↓

MARGARET SANGER

Margaret Sanger seated behind a desk and surrounded by twelve other women, New York, c.1924.

heredity.[282] The key premise of eugenics, a term coined by British statistician Francis Galton in 1883, is that "good" and "bad" qualities in people are inherited, in the same way that Gregor Mendel observed that the characteristics of plants are inherited in predictable ways from one generation to the next. According to historian Wendy Kline, eugenics offered a seemingly objective way to counteract what many saw as the troubling degradation of the "race" simply by controlling how selected "good" or "bad" traits were or were not inherited. It meant that the part of the population afflicted by these innate, genetic "bad" qualities could be eliminated in a few generations simply by preventing them from reproducing.[283]

For us today, eugenics often conjures the genocidal regime of Nazi Germany, but in the early twentieth century, it was a wildly popular ideology that seemed to be backed by cutting-edge science. Many eugenicists saw it as a way to "solve" the problem of increasing numbers of non-white immigrants and the "feeble-minded," a term that came to mean everything from people with disabilities to "loose" women. This application of the science is usually called "negative eugenics," and it involved discouraging undesirable people from having children or forcing them to be sterilized. "Positive eugenics" was about encouraging "fit" people (white, middle-class) to have more, healthier children to force a demographic shift. In the United States, eugenics was so popular and socially acceptable that "Fitter Family" competitions were common, in which white families were ranked by their desirable characteristics in a carnival-like atmosphere. It was from these applications of eugenics, as popular public programs for ensuring racial purity, that Nazi eugenicists took inspiration for their own eugenics programs.

In Britain, more so than in America, class was the most important gradient along which reformers applied eugenic theories. Influential British obstetrician and eugenicist Mary Scharlieb (June 18, 1845–November 21, 1930) conceived of eugenic measures this way. Scharlieb was trained in India where she was living with her husband, and at the School of Medicine for Women in London on their return to Britain. After postgraduate training in Vienna and another stint practicing in India, she received an MD in 1887. Historian Greta Jones shows that Scharlieb and other middle-class reformers interested in eugenics sought to repair the degeneration in the middle class evinced by their decreasing birthrate, while what they saw as immoral, libidinous overproduction among the poor was the chief symptom of their degeneration.[284]

But Scharlieb, like many other women involved in the eugenics movement, saw eugenics as potentially liberatory for women. During this time, the state took a positive eugenic, pro-natalist position by encouraging desirable middle-class families to have more children and halt the declining birth rate. Pro-natalist ideology had the potential to bring esteem to women as essential protectors of the race, but it also threatened to steer them away from the social freedoms that suffragists and others fought for. Proponents of pro-natalism hoped that by highlighting the eugenic imperative of motherhood, more equal marriages would emerge in which women were valued for their roles and relieved of some of the drudgery of domestic life.[285] Scharlieb was opposed to birth control because it compounded already low middle-class birth rates.

As in other instances where the looseness of eugenic ideology resulted in contradiction and conflict within the eugenic movement, the question of the use of birth control as a eugenic tool was constantly debated within the birth control movement. As one of Marie Stopes's contemporaries, Scharlieb was compared to Stopes in her own time, and observers noted their difference of opinion on the question of contraceptives. After Scharlieb's views were published in the journal *Nature* in a book review, Stopes wrote to the journal to argue that Scharlieb's views were religiously motivated and did not refute the safety and efficacy of the birth control methods, like the diaphragm, that Stopes advocated for on medical and scientific grounds.[286]

↑ A "Prorace" brand cervical cap from England, c.1915–1925.

In the United States, Margaret Sanger (September 14, 1879–September 6, 1966) is one of the most famous historical figures of the twentieth century, and as such, she is a common inclusion in lists of important women in science. But unlike Stopes, she was not herself a scientist. Born Margaret Louise Higgins in 1879 in Corning, New York, Sanger was educated at Claverack College and the Hudson River Institute before entering nursing training at White Plains Hospital in 1900. Two years later she married architect William Sanger, and they had three children, all within the first decade of the twentieth century. Not until 1911 did Sanger's radical political transformation begin, after the family moved from the leafy Westchester suburbs to New York City. Exposed to the bohemian society of the city, Sanger became friends with socialists and labor activists, and even participated in a Massachusetts textile workers strike in 1912.[287] During these early years in the city, Sanger worked as a visiting nurse and learned about socialism from none other than anarchist activist Emma Goldman herself. These were formative years for Sanger as she came to understand how the prevention of pregnancy could be liberatory for working class people.[288]

By the time she began publishing her periodical *The Woman Rebel* in 1914, Sanger had already run foul of various obscenity laws by writing a column for *Call* about sexual health and hygiene. For Sanger's career, the most important of these laws were the Comstock Laws of 1873. Sanger understood these laws to be a threat specifically to working women, who she said "have always been thrown into the hands of the incompetent" as a result of doctors being prohibited from providing information on family limitation.[289] As a socialist, her commitment to birth control was also a commitment to the liberation of the working class. Of *The Woman Rebel* she wrote, "the aim of this paper will be to stimulate working women to think for themselves and to build up a conscious fighting character."[290] It was in *The Woman Rebel* that Sanger coined the term "birth control." And it was because of that periodical that she was indicted for violating obscenity laws in 1914 and was forced to flee to Britain, where she met Stopes during her sojourn in London. When she returned to the United States Sanger's charges were dropped after her daughter died unexpectedly and the public pressed for sympathy.[291]

In 1916, Sanger opened the first birth control clinic in the United States modeled on Aletta Jacob's clinic (see page 148), which Sanger had visited during her time in Europe.[292] Located in Brooklyn, New York, the clinic distributed information about birth control but not the devices themselves, because there was no doctor on staff to fit patients with the appropriate-size diaphragms. Sanger believed she could get around the Comstock Laws by giving out information in person, rather than using the postal service, which was specifically prohibited by the law. But the clinic was raided by police and closed down only a week after opening, and Sanger and her staff were arrested.[293] Sanger's appeal was in its own way a victory. If the court ruled that had Sanger had a physician on staff, then the clinic could have operated for medical reasons. Sanger would use this ruling to model her later clinics to avoid future raids.[294]

Sanger began publishing a new periodical called the *Birth Control Review* in 1917. Around this time, the reach of Sanger's radical, socialist approach to advocating for birth control was too limited, and she sought out new audiences for her message. As historian Cathy Hajo points out, birth control was seen as extremely radical, much more so than the contemporaneous movement for eugenics, and Sangers ideas were met with considerable resistance in mainstream society. For non-scientists like Sanger, allying birth control with the scientific authority of eugenics, touted by famous academics, statesmen, and public figures, was an opportunity to bring prestige to a movement the public viewed as fringe.[295]

The fit between birth control and eugenics in the United States was far from seamless, and many of the contradictions within the British feminist movement also proliferated in America. As a method of negative eugenics, birth control was antithetical to the pro-natal positive eugenics directed at the middle classes encouraging them to have more children. Sanger and many other birth-control advocates understood that a person's social and economic status, which eugenicists viewed to be as a result of unchangeable genetic flaws, could in fact be changed—in some cases by simply limiting pregnancy.[296] Hajo points out that this type of birth-control activist did indeed hold similar beliefs to eugenicists about the superiority of white people, but their belief was more often a result of birth controllers' social and economic position, rather than adherence to eugenic ideologies.

Despite the contradictions and disagreements between the two movements, birth control and eugenics were intertwined, and Sanger was among those who used eugenic rhetoric to promote birth control, even if she did not do so consistently throughout her career. Her early work focused on the ways that birth control can alleviate the suffering of poverty through family limitation, acknowledging that it is deprivation that causes the environment in which someone lives to contribute significantly to their "fitness." It remains unclear whether or not Sanger believed in hard-line hereditary eugenics, which held that people from "inferior" races and ethnic groups should be restricted from reproducing, and this confusion is common when evaluating the position of

BIRTH CONTROL--WHAT IT WILL DO

It will give every mother the right to have children only when she feels that her health and strength will allow her to give them the care and attention they need.

It will enable her to arrange for proper intervals of time between her babies.

It will give her the possibility of recovering her strength in case she is worn out physically or nervously, or has any disease aggravated by pregnancy.

It will enable her to gain strength if she has worked hard and long hours before her marriage. No woman should become pregnant until she is well rested from fatiguing labor.

It will give her time to know her children, and to devote herself to bringing them up.

It will give her a chance to develop mother-love, instead of becoming a slave, a worn out, broken, spiritless drudge.

It will keep her husband's love and attention. Parents will have only the number of children they want, and at such intervals as will keep their interest alive and their love cemented by companion-ship and harmony.

It will prevent the practise of taking drugs and poisonous nostrums to avoid undesired pregnancy.

It will prevent the death of mothers whose physical strength cannot stand the strain of pregnancy.

It will prevent the death of thousands of babies whose passing out is caused by poverty, ignorance, neglect and insufficient vitality inherited from exhausted mothers.

It will prevent child labor. Poor mothers will be helped and advised to have only the children their husbands can support.

It will prevent prostitution:—because
    (a) Young people will be able to marry early and wait until their incomes are sufficient before having children.
    (b) Wives will be freed from the haunting fear of pregnancy which hovers over a woman from month to month, and fre-quently drives husbands to prostitutes.

It will prevent disease, especially the transmission of disease from parents to offspring.

It will set the woman free to show her affection and express her love for her husband, an expression which will hold husband and wife together.

It will make of the home a place of peace, harmony and love. The man will want to come to it; the woman will find in it her happiness and development; the children, well nurtured and carefully edu-cated will grow up in it to be the greatest assets of the nation.

Join the AMERICAN BIRTH CONTROL LEAGUE

MARGARET SANGER
President

104 Fifth Avenue
New York City

←

AMERICAN BIRTH CONTROL LEAGUE

A flyer for the American Birth Control League, the organization founded by Margaret Sanger in 1921.

↓

MARIE STOPES

British botanist and birth control advocate Marie Stopes was heavily involved in the birth control movement in addition to her successful career in science. Stopes was, like many scientists and public figures of her time, a eugenicist.

birth control advocates in this period.[297] But Sanger did employ the language and arguments of negative eugenics, especially when seeking support for the birth control movement from scientists and doctors who subscribed to eugenic ideology. In order to legitimize the radical proposition of birth control at this time, advocates needed to situate their message within the socially acceptable and scientifically authoritative reputation of eugenics.[298]

Sanger was an indefatigable campaigner for birth control almost all of her adult life. Through the 1920s, she founded the Birth Control Research Bureau and a legislative lobby to push for new laws on birth control. In 1937, the movement won an important victory in the *United States v. One Package of Japanese Pessaries* when a New York Court ruled that doctors had the right to supply patients with contraceptives for medical reasons. This effectively legalized clinics for birth control if they were overseen by doctors.[299] As birth control became more socially acceptable, Sanger's radical message and her lack of medical expertise was seen as a liability for the movement. She was forced out of the American Birth Control League in 1928, and in 1942, the League changed its name and those of its nationwide network of clinics to Planned Parenthood. Today, Sanger is perhaps even more famous for her support of Gregory Pincus's research on hormonal contraceptives, which led to the development of the first

↑ Prorace brand of spermicidal pessaries from England.

birth control pill, approved by the Food and Drug Administration in 1960. But this was only the culmination of her lifelong struggle for reproductive rights. Sanger died in 1966, having seen the movement she helped to build bear fruit through the twentieth century. She and Stopes lived during, and helped create, a pivotal time in the history of women's rights.

People like Sanger and Stopes led extraordinary lives, and the importance and resonance of their work is worth remembering and studying. But canonizing them as unalloyed heroes of the cause of reproductive freedom, or casting them as villains of the eugenic movement, obscures the more important questions of the way science molds and is itself shaped by the social and cultural norms of its time. Eugenics in the early twentieth century was considered cutting edge science with a positive practical application, and it was far less controversial than the ideas of reproductive freedom and family planning that the birth-control movement advocated. The mainstreaming of eugenics in Britain and America became a model for Nazi eugenics programs in the interwar period. The questions of whether or not any one birth control advocate was a eugenicist or if contraception is immoral or eugenics are much less important than understanding the complex social and cultural context of the history of birth control and the women who pioneered its use.

CHAPTER 14

Women Archaeologists and Anthropologists
Humanize their Past

CHANGING THE NARRATIVE

In 1928, Mexico City celebrated the Aztec New Year for the first time since its eradication by the Spanish Conquest in 1519. The Aztec New Year marks a solar phenomenon that occurs twice a year when the sun casts no shadows as it reaches the zenith of each latitude twenty degrees north and south of the Equator. For the ancient people of the Tropics, this phenomenon signaled the sun god's descent to earth, bringing with it a period of heavy rainfall to nourish crops. Earlier in 1928, Mexican-American archaeologist and anthropologist Zelia Nuttall wrote a spirited defense of the festival's renewal, writing that the accurate predictions for the solar year made by ancient Mexicans "is an honor to this race, and an intellectual contribution to the sum total of human wisdom which deserves world-wide appreciation for its originality and importance."[300]

The reinstatement of the festival was the culmination of Nuttall's career, which was dedicated to the recovery of ancient Mexican life. After centuries of salacious narratives of exaggerated Mexican savagery spun by European colonizers, Mexico's present-day people could take pride in their culture. The entry of women into Archaeology and Anthropology in the early twentieth century would signal a change in how the fields viewed the cultures of the past and the contemporary descendants of those cultures. By using the tools of their science, women like Nuttall attempted to redress the harmful narratives of colonization that had long dehumanized Indigenous people of the Americas.

← The *Codex Zouche-Nuttall* is a pre-Columbian document, produced between 1200 and 1521, of Mixtec pictography. Zelia Nuttall recovered the codex and published it with an introduction interpreting the content and detailing its historical context.

Born in San Francisco to an Irish father, Nuttall was more drawn to her Mexican mother's heritage and native home than her Euro-American roots, and archaeology gave her the special tools and insight to explore both. Late-nineteenth-century and early-twentieth-century archaeology was in many ways a Euro-American colonial and imperial enterprise. Western countries' claim to Indigenous ruins and their accumulation of artifacts and monuments had become a way for these nations to broadcast their imperial power around

By using the tools of their science, women like Nuttall attempted to redress the harmful narratives of colonization that had long dehumanized indigenous people of the Americas.

the globe.[301] This collecting of ruins and antiquities gave Euro-American archaeologists the opportunity to control not only the material history of these sites but also stories about their people and knowledge about their cultural past. Colonizers thus had the power to present an image of Indigenous people in whatever way served their purposes and further justified imperial domination.

It was in this fashion, Nuttall claimed, that the world had come to see ancient Mexicans as little more than "bloodthirsty savages, having nothing in common with civilized humanity." In a 1897 article, "Ancient Mexican Superstitions," Nuttall argued that stories of Aztec human sacrifice had taken "such a hold upon the imagination that it effaces all other knowledge about the ancient civilization of Mexico." This was a falsehood she laid at the feet of "Spanish writers" whose reports of the Aztec spiritual ritual were "grossly exaggerated, some say purposely, in order to justify, in the eyes of the civilized world, the cruel extermination of the native civilization."[302] Nuttall would attempt to change this story by correcting the archaeological record of ancient Mexico.

When Nuttall began to pursue Archaeology and Anthropology, it was, like most scientific fields, overwhelmingly white and male, but women were still involved in excavation and fieldwork in manifold ways. White women participated as wives to male scientists, secretaries, or assistants or in museums as researchers and cataloguers, while Indigenous women were often informants or interlocutors for anthropologists and ethnographers. Even women who spearheaded their own excavations and research did not usually receive a PhD or even any formal university training and were supported in large part by male archaeologists and wealthy patrons. Without formal titles or degrees, many of these women were "amateurs" by default and have slipped through the cracks of in the history of archaeology and anthropology. But as many as two-hundred women have been identified in Americanist archaeology between 1865 and 1940. Nuttall was one of them.[303]

Nuttall published her first paper, "The Terracotta Heads of Teotihuacan," in *The American Journal of Archaeology and the History of Fine Arts* in 1886. Two years before, she had visited the historic site of Teotihuacan, located northeast of modern-day Mexico City, with her mother, brother, and daughter, Nadine, whose father Alphonse Pinart she had recently separated from. There she collected a series a small terracotta heads, and she initiated a comparative study of her collection with other undated heads whose cultural significance was not fully understood. Nuttall was able to date the origin of the artifacts to near the time of the Spanish Conquest and credited their creation to the Aztecs, for whom they were representations of the dead. This study quickly caught the attention of Frederic W. Putnam, the director of the Peabody Museum in Massachusetts. Putnam lauded Nuttall in the Museum's 1886 annual report, writing

> "[she is] familiar with the Nahuatl language...with an exceptional talent for linguistics and archaeology. As well as being thoroughly informed in all early native and Spanish writings relating to Mexico and its people, Mrs. Nuttall enters the study with a preparation as remarkable as it is exceptional."[304]

That same year Putnam made Nuttall an honorary special assistant in Mexican archaeology at the Peabody.

———————

Nuttall's merely honorary position at the Peabody, and her "amateur" career offered her more freedom to pursue what was of interest to her than a job at a museum would have. With her independent familial wealth and the financial patronage of newspaper heiress Phoebe Hearst, Nuttall took her study abroad for the next thirteen years. Researching, studying, and collecting artifacts and manuscripts, she visited libraries and collections around the world, from Morocco to Russia, but what was ultimately of most interest to Nuttall was the subject of her first paper—Mexico. Across more than a dozen publications spanning her career, she analyzed ancient Mexican calendar systems and astronomical methods; collected and interpreted ancient folklore and rituals; and recovered, interpreted, and published Aztec codices that had been left to fall into obscurity in dusty European collections.

UNITING THE PAST AND THE PRESENT

Nuttall often found herself on contentious ground because Mexican archaeology was deeply embroiled in its national politics in the early twentieth century. While politicians and intellectuals thought Mexico's history of Indigenous empires lent the country an unrivaled national distinction, they also wanted to leave the past on display on shelves in museums, where it could be admired from a distance. They rejected any connection between present-day Mexicans and the supposed "savage" Aztecs of the past. They worried that identifying with such history made modern Mexico appear backward and uncivilized to rest of the modern world.[305] The question of modern Mexicans' relationship to their Aztec ancestors was at the heart of Mexican archaeology, and consequently, Nuttall's career.

Nuttall took an unequivocal stand on this controversial question, arguing that "the Aztec race is represented by thousands of individuals, endowed with fine physiques and intelligence, who speak, with more or less purity, the language of Montezuma."[306] She argued that the popular portrayal of ancient Mexicans as uncivilized prevented modern Mexicans from taking pride in this Indigenous heritage. Her work collecting and sharing the rich history of Mexico's past, she hoped, would "lead to a growing recognition of the bonds of universal brotherhood which would unite the present inhabitants of this great and ancient continent to their not unworthy predecessors..."[307]

When Mexico City ultimately reinstated the ancient Aztec New Year in 1928, Nuttall could not help but see the event as a victory, for both her scientific work and for Mexico. Elated by the celebrations, she wrote to a friend, marveling at what archaeology could offer a culture: "It is strange to have archaeology produce such lively offspring! You can imagine how happy it has made me to have extracted from the grave of the past a germ so vital and lively that it will set children dancing and singing and observing the sun every year."[308] Nuttall died five years later in Casa Alvarado. The house is now the location of the National Sound Archive, a collection of Mexico's musical heritage.

What set Nuttall apart from other archaeologists was not her Mexican descent, but her insistence on an archaeology for Mexicans. Throughout her career, she advocated for the restoration of ancient rituals, supported the careers of Mexican archaeologists, and prioritized what was important to Mexicans. For Nuttall, archaeology was a window into the past of the country she loved above all others, but it was also a way for her to take part in the politics of Mexico's present and in the preservation of its culture for the future.

SALVAGING NATIVE AMERICAN LIFE

While Nuttall was winding down her career and celebrating her achievements in Mexico, Bertha Parker (August 30, 1907–October 8, 1978), also known by her Seneca name Yewas, was just beginning her archaeological career in the United States. Parker was born at an archaeological site in Silverheels in New York in 1907 where her father Arthur Caswell Parker and his wife Beulah Tahamont were working at one of his excavations. Arthur was a promising archaeologist and anthropologist of Native American culture of mixed European and Seneca ancestry, and Beulah was daughter to the chief of the Abenaki, a First Nations band government belonging to the Eastern Algonquian people in Canada.

Arthur and later Bertha were part of an important shift in archaeology and anthropology. Largely influenced by the famous anthropologist Franz Boas, early-twentieth-century anthropology was entering its "salvage era." Anthropologists and institutions launched expeditions to communities that Boas and his cohort deemed on the brink of extinction or complete assimilation by Anglo-American culture. Assimilation was the central policy of the American mission of civilizing its First Nations people through forced re-education, relocation, and often sterilization; anthropologists did not want to miss their chance to study the "authentic Indian" before they were gone.[309] These expeditions to Native communities often resulted in a myopic representation of Native people, one which reinforced settler beliefs of an inferior primitive people and a superior Euro-American people.[310]

Native Americans often served as unnamed informants for white anthropologists and had very little say in how their stories and culture were represented and then codified in scientific writing. Thus, the entry of Native American archaeologists and anthropologists was an important corrective during the salvage era and beyond, but the field itself remained entrenched in

←

BERTHA PARKER

Bertha Parker was the first professional Native American female archaeologist and the first Native American woman to work as an ethnologist and archaeologist for a major museum.

↓

BERTHA PARKER WITH HER DAUGHTER

Bertha Parker, her daughter Billie and Johns H. Harrington looking at Spanish dishes at the Casa de Adobe in Los Angeles, California in 1929.

colonial models. Arthur studied the Iroquois and Seneca people and culture, but despite being of Seneca descent, he demonstrated a disdain for it too. He disliked traditional dress, opting instead for a suit and tie, and often mocked tribal superstitions and stories. And despite protestations from the Seneca, he conducted excavations on tribal lands.[311] Bertha, however, would do things differently.

Bertha did not immediately take up her father's mantle in the field. During her teen years and early twenties, she lived in Hollywood with her actor grandparents, the Tahamonts, and her mother, who had divorced Arthur by this time. In California, she met Yuman actor Joseph Pallan, and they were wed shortly after Bertha became pregnant, but the marriage was not a happy one. Pallan was abusive, and when Bertha pursued divorce, Pallan kidnapped her and their daughter Wilma Mae, holding them in a brothel in Mexico. Anthropologist Mark Raymond Harrington, a colleague of Arthur's and once Beulah's partner, came to her rescue. Harrington was married to Bertha's aunt Edeka, Arthur's sister. From Mexico, Harrington brought Bertha and Wilma Mae back to his archaeological site in the desert of Nevada, where Bertha stayed and joined Harrington's crew.[312]

SETTING A NEW PRECEDENT

Starting as camp cook and secretary and rising to assistant archaeologist, Bertha became an indispensable member of Harrington's team. She learned everything on site, where Harrington encouraged an atmosphere of collaboration and equality in which women and men and Native Americans and white people worked and learned alongside one another. In the evenings, Harrington held education sessions for the crew, lecturing on anthropological and archaeological theories and methods.[313] In the mornings, Bertha cooked for the camp and then traded the spatula and frying-pan for a trowel and headlamp in the afternoons to explore and dig on her own. Harrington described Bertha's archaeological work as a sort of longing:

> *"I suspect she wishes the afternoons were longer, so she could put in more time digging in the ruin on the mesa. If you want to see her black eyes shine talk to her about archaeology, or watch her uncovering something choice."*[314]

Bertha did indeed uncover several "choice" items while working on the crew. The first was a Pueblo site, which she named Scorpion Hill in 1929.

She conducted a solo excavation of the site and later published the details in a 1933 article in *Masterkey*, the journal for the Southwest Museum of the American Indian, which financed Harrington's expeditions. She later exhibited photographs and artifacts from the dig at the Southwest Museum.[315] The next year, Bertha drew the attention of the international archaeological community when she discovered the skull of an extinct giant ground sloth from the Pleistocene era, the *Nothrotherium shastense*, in Gypsum Cave in Nevada. Bertha made the discovery on one of her usual afternoon excursions, which Harrington later described in *Desert Magazine*:

> "[A]s was her custom, when the paper work was done, [she] donned headlight and dusk mask and proceeded to the cave to search the crevises[...] There she spied a curious object that looked like a bone. Extracting it with difficulty she found it to be the skull of a strange animal, unlike anything she had ever seen."[316]

What really fascinated the crew and archaeological community were the human artifacts—yucca fiber twine and weapon fragments—she found in a layer beneath the sloth skull.[317] Finding human artifacts near the remains of the sloth prompted new questions about when humans arrived in North America, and the mystery surrounding human origins invited the support from the California Institute of Technology. Unfortunately, one of the essential members of the crew and Bertha's new husband, James Thurston, died unexpectedly on site, and the excavation ceased.

From 1931 to 1941, Bertha worked as Assistant in Archaeology and Ethnology at the Southwest Museum in Los Angeles, making her the first professional Native American woman archaeologist and the first Native American woman to work as an ethnologist and archaeologist for a major museum.[318] Bertha began a series of trips to Native communities during which she interviewed Maidu medicine men, Paiute basket makers, Pomo bear maskers, and storytellers from the Maidu and Yurok tribes, all of which she subsequently published in *Masterkey*. Unlike her father, Bertha honored tribal protocols and rituals. Breaking with tradition in anthropology, she often named her female informants when appropriate and even went so far as naming some as co-authors on published pieces.[319] She gave Native informants visibility and a level of control over how their stories were presented to the scientific community, when other anthropologists built entire careers on the knowledge and stories of their nameless subjects.

Although she resigned her position in 1942 when her daughter died in a hunting accident, Bertha remained on the board of trustees for the museum. She also participated more actively in the Native Rights Movement in California with her third husband, Espera Oscar de Corti, an Italian actor who adopted a Native American persona by using the name Iron Eyes Cody in films. Even though de Corti supported Bertha's archaeological and anthropological work and himself advocated for Native Americans in Hollywood, he is part of the reason it has taken so long for Bertha to be recognized for her work. After she died, de Corti represented her as a heavy-drinking partier in his autobiography, *Iron Eyes Cody: My Life as a Hollywood Indian*, and claimed that he was part of Bertha's discoveries at Gypsum Cave and her anthropological work in Native communities.[320] Even in death he maintained some control over her story with her gravestone simply reading "Mrs. Iron Eyes Cody."

***

Neither Nuttall nor Parker received professional training in archaeology and anthropology, but the value of their discoveries cannot be denied. Perhaps more valuable is the humanizing perspective they brought to their subjects in a scientific field rife with racist beliefs about the Indigenous peoples of the Americas. In a time when Mexicans were often seen as salacious savage stereotypes, Nuttall challenged the world to see them as so much more. Parker allowed her Native American informants to be named and put them on equal footing as her co-authors instead of letting them fall into anonymity. Both scientists understood that their investigation of the past had real consequences for the living and future descendants who live on its ruins.

Zelia Nuttall

(September 6, 1857–April 12, 1933)

Born in San Francisco, Zelia Maria Magdelena Nuttall was an anthropologist and archaeologist who specialized in Aztec Mexican cultures and pre-Columbian manuscripts. Nuttall was the second of six children to an Irish father, Robert Kennedy Nuttall, and a Mexican-American mother, Magdalena Parrott. She spent many of her formative years traveling in Europe and received her first formal education at Bedford College in the United Kingdom.

In 1884, she undertook her first archaeological study at the historical site of Teotihuacan in Mexico. She conducted a comparative study of Aztec terracotta heads, and after her 1886 paper on the topic, she was made an honorary special assistant in Mexican archaeology at Harvard's Peabody Museum, a position she held for forty-seven years. In 1887 she was appointed as a fellow in the American Association for the Advancement of Science.

Among her most important finds are two pre-Columbian manuscripts of a pictographic history of ancient Mexico: the *Codex Zouche-Nuttall* (1902), which she recovered from a private library in England, and *Codex Magliabechiano* (1903), recovered from a library in Florence. She also published *The Fundamental Principles of New and Old World Civilizations* (1901), *The Book of Life of the Ancient Mexicans* (1903), and *New Light on Drake: Documents Relating to His Voyage of Circumnavigation, 1577–1580* (1914). In 1910, she initiated an excavation on Isla de Sacrificios where she found the ruins of a site for human sacrifice.

In 1905, she moved to Mexico and purchased a sixteenth-century mansion, which she named Casa Alvarado. She lived in Mexico until her death in 1933.

NOTICE
YOU MUST WEAR
YOUR BADGE IN
PLAIN SIGHT IN
THIS AR

CHAPTER 15

What Cannot Be Unmade

GOING CRITICAL

On the afternoon on December 2, 1942, Leona Woods (August 9, 1919–
November 10, 1986), a graduate student at the University of Chicago kept
a careful, trained eye on her boron trifluoride neutron detector, as one by
one the cadmium-plated control rods were removed from the twenty-foot
high pile of atomic material. The cadmium control rods would absorb rogue
neutrons until Woods and her colleagues were ready to unleash them into the
radioactive uranium. One by one, the rods were removed. Once the nucleus of
the radioactive uranium takes on an extra neutron, the nucleus splits into two,
releasing a massive amount of electromagnetic radiation, radioactive fragments,
and even more neutrons. These neutrons in turn collide and absorb into another
uranium nucleus, splitting it once again and releasing more radioactive energy.
Woods watched as her neutron detector registered the first occurrence of
nuclear fission in the pile while calling out the detector's readings. Enrico Fermi
ordered another control rod be removed; Woods shouted more readings. The
pattern—splitting, energy release, splitting, energy release—repeated until the
pile went critical. There, under the University of Chicago's Stagg Field football
stands, on what used to be a squash court, the world's first human-made self-
sustaining nuclear chain reaction was created.

The path to the breakthrough in Fermi's lab was charted just a few years
before, when in January and February of 1939, two major papers on nuclear
physics were published Europe. The first, written by German chemist Otto
Hahn, provided evidence for what could happen when the nucleus of uranium
absorbs an extra neutron. The second, written by Austrian physicist Lise

← Manhattan Project workers at the Oak Ridge site in Tennessee.

Meitner (see page 248) and her nephew Otto Frisch, gave a physical explanation of this phenomena and named it "fission." These papers provided the fundamental building blocks for an atomic bomb, and physicists in the United States knew it. A few months after the two papers appeared, physicists Albert Einstein and Leo Szilard wrote a letter to US President Roosevelt alerting him to the possibility that Germany was building such a bomb. The United States resolved to get there first.

The Manhattan Project, a massive, highly secretive mobilization of scientists and laboratories tasked with creating the weapon, did not launch right away. Roosevelt first formed the Advisory Committee on Uranium, a group of military and scientific experts tasked with determining and reporting on the feasibility of a nuclear chain reaction. In its first report from November 1, 1939, the committee advised the president to support research and development into uranium oxide and also to fund Fermi and Szilard's experiments with the atomic pile (later called a reactor) at Columbia University.[321] On December 7, 1941, the United States officially entered World War II when Japan attacked

While Woods was the only woman to work on the first chain reaction, hundreds more worked on the Manhattan Project itself.

Pearl Harbor, and the United States saw in the destructive power of the bomb the potential of a decisive and quick Allied victory. Fermi moved his operation from Columbia to the Metallurgical Laboratory at Chicago in February of 1942, and then on August 13, 1942, the Manhattan Project became official. When Meitner, whose research had proved the weapon was feasible, was asked to join the Manhattan Project, she vehemently refused, famously saying "I will have nothing to do with a bomb!"

When the Chicago atomic pile went critical, the possibility of a nuclear chain reaction passed from theory to reality. Forty-seven people were at the Metallurgical Lab to witness the historic event, and Leona Woods was the only woman among them. As a doctoral student in molecular spectroscopy and one of the youngest scientists on the team, her role was to monitor the neutron activity in the pile using the detector that she herself had constructed. Before everyone left the lab that night, Fermi passed around a bottle of Chianti to celebrate. The team signed the bottle to commemorate their achievement, and their hopes for a new technology they believed could end the war. As everyone silently sipped their wine out of paper cups, Woods voiced what was undoubtedly on the minds of many in that room: "Let's hope that we are the first to succeed."[322]

WOMEN OF THE MANHATTAN PROJECT

While Woods was the only woman to work on the first chain reaction, hundreds more worked on the Manhattan Project itself. Massive in scale and scope, the Project had sites across the United States, each working on several aspects of the development and construction of the atomic bomb. J. Robert Oppenheimer oversaw "Site Y" in Los Alamos, New Mexico, where the bombs Fat Man and Little Boy were made. Over a thousand miles to the east in Oak Ridge, Tennessee, "Site X" was home to three uranium enrichment plants, which separated the fissile isotope uranium U235 from non-fissile uranium metal U238. Northwest in Hanford, Washington the first production-scale plutonium reactor transmuted irradiated uranium into plutonium. In addition to the three main sites, universities and laboratories affiliated with the Project operated across the nation.

In the Tech Area of Los Alamos, the most well-known of all the sites, an estimated thirty percent of the labor force were women, totaling about two-hundred employees by 1944. Diverse in expertise, they worked in several areas of bomb development with rough numbers showing twenty-four women working in chemistry and metallurgy, twenty in bomb engineering, eight in ordinance, four in experimental physics, and another four in explosives.[323] Los Alamos also took advantage of a workforce of women computers, who calculated the behavior of the bombs' implosion, including the incoming and outgoing shock wave of the explosion.[324] Many of these computers were married to Los Alamos scientists who lived and worked on site. Los Alamos also engaged the efforts of the ENIAC, the first electronic general purpose computer, and its female programmers in Philadelphia to test different models for creating a thermonuclear explosion. Theoretical physicists Stanley Frankel and Nicholas Metropolis wrote a program for one such mathematical model. Using some one million IBM punch cards, the women programmed Frankel's

↑

CHIEN-SHIUNG WU

Chien-Shiung Wu, an experimental physicist,
worked on the Manhattan Project out of Columbia
University and helped Enrico Fermi reach a
sustained chain reaction in plutonium reactors.

and Metropolis's model into ENIAC and delivered the results pointing to flaws in their bomb design.[325]

At Hanford, women made up about nine percent of the 51,000 employees in 1944 when the staff was at its largest.[326] Leona Woods and her husband, John Marshall, moved to the Hanford site to oversee production of the plutonium reactors. One of the most crucial developments at Hanford was made by Chien-Shiung Wu (May 31, 1912–February 16, 1997), an experimental physicist based at Columbia University. An immigrant from Shanghai, Wu studied the nuclear interactions of noble gases, and when the Hanford plutonium reactors failed to sustain a chain reaction, Fermi called on Wu to diagnose the problem. After reviewing the Hanford data, Wu determined that the presence of the xenon isotope Xe-137 was absorbing neutrons, causing the reaction to shut down on its own. Using Wu's data, Fermi was able to compensate for the xenon by adding more atomic material. Her work also led to developments in controlled shut-down and restart mechanisms for nuclear reactors—by controlling the flow of xenon gas in and out of a reactor, operators could quickly shutdown a reactor when needed.[327] Wu never even visited the Hanford site to make her conclusions, but she continued to contribute to the Project at Columbia with a group working on the separation of U-235 from U-238.

Of all the Manhattan Project sites, Oak Ridge, Tennessee, was the largest, with a booming population of around 70,000 people at its peak. Most of the workers were women: statisticians, chemists, technicians, and control panel operators for uranium enrichment plants as well as computers and secretaries. These women came from a variety of backgrounds, some with college degrees or scientific expertise and some with a high school diploma or civil service exam qualification. Whether college-educated not, there was always work that needed to be done at Oak Ridge with three uranium processing plants and a pilot plutonium reactor on the site.

Not all the work at the three major sites was scientific or technical. Because these large sites functioned much like company towns where families lived and children grew up, they needed to fill roles that supported the overarching social and civic infrastructure of the town. There were teachers, nurses, and sanitary workers at all sites. Oak Ridge even had its own newspaper, the *Oak Ridge Journal*, with a staff of journalists. The wives of male scientists were encouraged to go to work for the war effort in any needed roles, which meant that those who were scientists themselves did not always get job that matched their expertise or skill. Director of the Manhattan

Project, General Leslie Groves, explained how the sites were built with the labor of women in mind. "The system was designed to encourage the wives of our people to work on the project, for those who worked obtained priority on house-hold assistance" he wrote in his 1962 memoir *Now It Can Be Told*:

> *"Some of the wives were scientists in their own right, and they, of course, were in great demand, but with labor at a premium we could put to good use everyone we could get, whether as secretaries or as technical assistants or as teachers in the public school that we started for children."*[328]

At Los Alamos, working wives were not offered promotions or paid according to prior work experience, but they were nevertheless expected to put in forty-eight-hour work weeks. As one working wife put it in the oral history *Standing By and Making Do: Women of Wartime Los Alamos*,

> *"The Working wife's salary, which was set very arbitrarily, was influenced less by her previous work history than by the fact that she really had no bargaining power. She lived, after all, in a sort of town community."*[329]

At Oak Ridge, only men as "heads of households" were offered onsite family housing, and unmarried women or married women with husbands working elsewhere were never considered heads of house, which meant commuting and additional financial strain for some women.[330] In these quickly thrown-together makeshift towns, living was rough. Residents constantly faced housing and water shortages, power outages, and limited resources.

Regardless of the hardship, many women who had male relatives fighting overseas were happy to contribute to the war effort in whatever way they could. Reports of Nazi genocide compelled them to seek an end to the war, even if that meant working on weapons research and development. The work was so secretive and heavily compartmentalized that the women turning knobs on the control panels for the uranium enrichment plants had no idea what they were actually turning knobs for. Wives were kept in the dark by their scientist husbands who had high enough security clearance to know the full scheme of the Project. For the majority of the people employed by the Manhattan Project, the true nature of their work was kept from them until August 6, 1945. That day, the first atomic bomb, the uranium-based Little Boy, was dropped on Hiroshima. It was followed by the plutonium-based Fat Man on Nagasaki three days later.

In the days and months that followed the bombings, especially after reports began to surface revealing the extent of the devastation in Japan and the lingering effects of radiation on human bodies, the workers dealt with the reality of their contribution in numerous ways. Some felt pure pride or horror in what they had done, while others struggled with a conflicted mix of shame and pride, guilt and joy that the war had finally ended in an Allied victory. The first atomic bombing might have heralded the end of the war, but it was only the beginning of a new age, and eventually a new war. The secrets of the atomic nucleus could no longer be hidden by security clearances and concealed in private military towns. The Cold War would embroil nations all over the world in nuclear tension and conflict for decades to come. What had been made could not suddenly be unmade.

The proliferation and testing of nuclear weapons continued into the postwar period, even as researchers were still investigating the effects of radiation on the planet and on people. Shortly after the war's end, the major Allied Powers—the United States, Great Britain, and the Soviet Union—went their separate ways in pursuit of nuclear energy and weapons. The United States, far ahead everyone else, established the Atomic Energy Commission (AEC), which took over the infrastructure of the Manhattan Project to continue the nuclear research and development. By 1948, both Britain and the Soviet Union had achieved their own nuclear chain reactions, and in the early 1950s, both had initiated their own hydrogen bomb projects.

Between 1952 and 1958, the three countries had carried out 223 nuclear weapons tests in the atmosphere, on the ground, and underwater, with the United States alone conducting more than 120 tests in the Pacific and Nevada.[331] The long-term health outcomes of nuclear weapons testing were still largely unknown, and any concerns raised by nuclear disarmament advocates and scientists about the effects of nuclear fallout on humans were quickly dismissed and downplayed by the AEC.[332] But the United States and the AEC were unable to control information about nuclear fallout for long. On March 1, 1954, the United States military detonated a thermonuclear weapon, codenamed Castle Bravo, over Bikini Atoll, a United States territory 2,300 miles southwest of Japan. The radioactive plume from the explosion traveled outside the predicted area, carrying its fallout to the inhabited Marshall Islands. Also affected was a Japanese tuna fishing boat, the *Daigo Fukuryū Maru*—every one of its twenty-three member crew suffered from acute radiation syndrome in the aftermath and one man died.

The event was felt across Japan, for even in peacetime, the country was once again grappling with the devastating effects of nuclear weapons. With so many nuclear weapons being tested in the ocean surrounding their country, Japanese scientists began studying in earnest how radioactive material could be circulated in the ocean. Following the bombing of Bikini Atoll, Katsuko Saruhashi (see page 181), a Japanese geochemist at Tokyo's Meteorological Research Institute (now called Japan Meteorological Agency), and her colleagues began studying ocean-borne nuclear contamination. Saruhashi and her team tracked the circulation of radioactive fallout from Bikini Atoll, finding that ocean currents pushed water contaminated with the radioactive isotopes Cesium 137 and Strontium 90 from the weapons' test site northwest to Japan. Because contaminated water was spread by ocean current patterns, nuclear fallout from weapons testing in the Pacific was not distributed equally. The western Pacific surrounding Japan showed a higher concentration of radioactive isotopes than the eastern Pacific on the coast of California. Saruhashi and her team published their findings in a paper titled "Cesium 137 and Strontium 90 in Sea Water" in the *Journal of Radiation Research* in June of 1961. The findings did not please the AEC in the United States.

The AEC decided to put Saruhashi's results to the test in 1962, inviting her to participate in a six-month-long analysis comparing her methodology and results to those of American scientists at Scripps Institute of Oceanography at

This nuclear age in which we now live was brought to bear in part by powerful world leaders and scientists but also by the women who pulled the sometimes literal levers of nuclear science.

the University of California San Diego. Saruhashi's mentor and collaborator on her original study, Yasuo Miyake, supported her research and encouraged her to go. In June of 1962, Saruhashi traveled to California to begin her comparative research. The Scripps team was led by Theodore Folsom, who implemented the Institute's methodology for studying fallout in the ocean. Saruhashi was, however, not on placed equal footing with the other scientists at Scripps. First, Folsom informed Saruhashi that she need not commute to the Institute every day and was given what amounted to a wooden shack to conduct her research.[333] She was given samples with twenty percent less concentration of Cesium than those given to Folsom, making her analysis more difficult.[334] After six months, the results of the analysis showed only a ten percent discrepancy between the two teams. The findings were published in March of 1963 in the paper, "A Comparison of Analytical Techniques Used for Determination of Fallout Cesium in Sea Water for Oceanographic Purpose." The results validated Saruhashi's original study, and the AEC was forced to accept her conclusions about nuclear fallout.

Saruhashi and Folsom's study of ocean-borne nuclear fallout could not be ignored. Combined with radioactive rainfall in Japan resulting from a Soviet hydrogen bomb test, world leaders had to confront the environmental consequences of using nuclear weapons. In August of 1963, nearly ten years after the Bikini Atoll disaster, representatives of the United States, Soviet Union, and the United Kingdom met in Moscow to sign the Treaty Banning Nuclear Weapons Tests in the Atmosphere, in Outer Space and Underwater, which was later signed by an additional 123 nations. Even though the treaty still allowed for underground detonations, it marked an important step for the movement for international nuclear nonproliferation.

When Leona Woods facilitated the first sustained nuclear chain reaction in 1942, she probably had no idea that she was witnessing one of the most significant moments in global human history. This nuclear age in which we now live was brought about in part by powerful world leaders and scientists but also by the women who pulled the sometimes literal levers of nuclear science. As human computers, chemists, or technical workers, women were as much a part of the Manhattan Project as the men whose names we have come to readily recognize. They too have become the architects of our present era in which the immense power of the tiny atomic nucleus holds the potential to impact us all.

Katsuko Saruhashi

(March 22, 1920–September 29, 2007)

Geochemist Katsuko Saruhashi was born in Tokyo on March, 22 1920. At the age of twenty-one, Saruhashi quit her job at an insurance firm to become a chemistry student at the Imperial Women's College of Science in the city. After graduating in 1943, she accepted a position at the Meteorological Research Institute, and while there, she worked towards her PhD in chemistry at the University of Tokyo.

As a member of the Geochemistry Laboratory at the Meteorological Research Institute, Saruhashi studied carbon dioxide levels in seawater. She developed Saruhashi's Table, a method for measuring CO_2 using pH, temperature, and chlorinity, which has become a global standard. She also discovered that the Pacific Ocean releases more CO_2 than it absorbs.

Saruhashi broke new ground in her study of ocean-borne nuclear contamination following the nuclear weapons test the United States undertook on Bikini Atoll in the Marshall Islands. Saruhashi discovered that nuclear contamination traveled along ocean currents, and in the case of the Bikini Atoll bombing, the fallout spread clockwise, northwest toward Japan.

The Atomic Energy Commission in the United States initiated a comparative study between Saruhashi's methodology and American scientists at Scripps Institute of Oceanography in San Diego. The resulting paper showed that Saruhashi's methods and conclusions were sound. Saruhashi's research played an important role in limiting nuclear proliferation around the world, thanks to the signing of the 1963 treaty.

Saruhashi became the first woman elected to the Science Council of Japan, the first woman to win the Miyake Prize for Geochemistry, and the first woman recipient of an award from The Society of Sea Water Science in Japan. In 1981, she founded the Saruhashi Prize, a prize awarded annually to a female role model in science. Saruhashi died of pneumonia in Tokyo in 2007.

Section V

The Twentieth Century, Post-World War II

CHAPTER 16

The Plight of Women Refugee Scientists
Coming to America

On September 1, 1939, Germany invaded Poland, laying siege to Poland's Military Transit Depot on the Westerplatte peninsula. What became known as the Battle of Westerplatte marked the official launch of World War II. But before open warfare broke out, Jewish people in Germany had already been fighting anti-Semitic oppression for years. Beginning in the early 1930s, the Nazi political party gained increasing power in national politics, culminating in Adolf Hitler's appointment as Chancellor in 1933. The Nazi party enacted swift and sweeping legislation to limit the civil rights of the Jewish people, ultimately leading to the concentration and extermination camps of the Holocaust. This had devastating implications for Jewish scientists, especially after the Nazi party passed the Law for the Restoration of the Professional Civil Service Act in 1933, which prohibited Jews and other so-called "non-Aryans" from holding jobs in government institutions including German universities. As a result, many scientists lost their jobs. As they watched their freedoms being stripped away one by one, more and more Jews sought escape to other European countries and to the United States. Across the Atlantic Ocean and farther away from the Nazi regime, the United States was the first choice of resettlement for many refugees.

Finding refuge in the United States, however, was not guaranteed. The Immigration Act of 1924 limited the number of immigrants to 150,000 per year and set quotas for different countries. Larger quotas were reserved for Great Britain and other Western European nations while smaller quotas were set for Southern and Eastern European countries, curtailing the number of Jews and

← Immigrants approach New York City c.1910.

Italians allowed in the country. Even as Nazi violence escalated in Europe, the United States did not increase their quotas. Clergy, scientists, and other scholars had the option to apply for a non-quota visa, under Section 4-D provision in the Immigration Act. Scientists who held a professorship in their home country and had secured another position in the United States were eligible as were those who already had relatives in the country. But for women scientists who still faced gender barriers in their professions, securing a position at an American college or university was more difficult than it was for men.

Immediately following the passage of the Civil Service Act in Germany, 12,000 scholars were banned from their profession in Germany. Nearly half of them applied for aid from the Institute of International Education in New York, which established an Emergency Committee in Aid of Displaced German Scholars to help place émigré scholars in American universities and colleges. The Emergency Committee promised they would not place a financial burden on universities. By securing financial aid from the Rockefeller Foundation and other philanthropic organizations, they were able to provide annual grants to pay the salaries of émigrés.[335] Initially, the Emergency Committee fronted $2,000 for each salary grant, which was matched by the Rockefeller Foundation, but they eventually decreased the amount to $1,000 as more and more Jewish scholars were displaced from their home countries. The Committee was only able to fund around 330 scholars, and only four of the eighty women scientists and mathematicians who applied were granted aid.[336]

THREE WOMEN REFUGEES COME TO AMERICA

Emmy Noether (March 23, 1882–April 14, 1935) was the first woman refugee to receive aid from the Emergency Committee after she was dismissed from the University of Göttingen, along with five other Jewish colleagues. Noether had been at Göttingen since 1915, when she was invited to join the faculty by David Hilbert, one of the most influential mathematicians of the early twentieth century. When Noether joined Göttingen, the male faculty opposed her appointment, so Hilbert was forced to take her on as an unpaid guest lecturer. But her influence on Göttingen and on the broader field of mathematics, and eventually physics, was profoundly felt.

Before taking her post at Göttingen, Noether had already made a name for herself at the University of Erlangen as the first woman in Germany to obtain a PhD in mathematics in 1907. Noether's specialty was abstract algebra, which deals with structures such as rings, groups, and fields instead of real

MARION EDWARDS PARK

←

Marion Edwards Park was the third president of Bryn Mawr College and early member of the Emergency Committee of Displaced German Scholars.

↓

EMMY NOETHER

Emmy Noether developed Noether's Theorem and is considered one of the most significant mathematicians of the twentieth century.

number systems. Her early work focused on a branch of abstract algebra called invariant theory, which deals with polynomial functions that do not change. At Göttingen, Noether set to work applying algebraic invariants to Einstein's Theory of General Relativity, which resulted in a discovery of her own. In what would become known as Noether's Theorem, she proved that for every symmetry in a physical system there is a corresponding conservation law. Her first theorem demonstrates that energy can neither be created or destroyed in a system that displays time symmetry. She linked the seemingly separate concepts of time symmetry to energy, translational and rotational symmetry to momentum, algebra, and physics, and in so doing opened a whole new world that mathematicians and physicists are still exploring.

When Germany banned Noether from the profession in 1933, after nearly twenty years at Göttingen, other mathematicians, including Hermann Weyl, petitioned the university over her dismissal. The university's chief administrator, J. T. Valentiner would not support her, citing Noether's Marxist politics as evidence that she did not support the German state.[337] As a Jewish woman with liberal politics, she had no chance of retaining her position under the Nazis' fascist regime.

Colleagues in the United States immediately went to work trying to secure Noether a position in an American institution. At Princeton, Weyl tried to arrange a position for her, and although Princeton participated in the Emergency Committee's efforts and welcomed Weyl and Einstein, the university would not take a woman. Princeton was "a men's university, which admits nothing female," Noether later wrote.[338] This was the case for many American ivy league universities, so women refugees often turned to smaller women's colleges like Bryn Mawr College in Pennsylvania. Solomon Lefschetz, a colleague of Noether's from Göttingen and then at Princeton, reached out to Bryn Mawr on her behalf. In coordination with the Emergency Committee and the Rockefeller Foundation, the college agreed to accept Noether with a salary of $4,000 for the academic year of 1933–1934, with the possibility of renewal.[339] Noether accepted, and on November 7, 1933, she arrived aboard the ship *Bremen* to take up her post.

Less than a year into Noether's tenure at Bryn Mawr, the college set up a scholarship in her name. Noether's students loved her, and in return, she loved them and the college that had embraced her. But before Noether could completely settle into her new life and career in the United States, she died suddenly on April 14, 1935 from circulatory collapse following surgery to remove an ovarian cyst.

Following Noether's death, the Bryn Mawr community, her students, and the mathematics community at large expressed an outpouring of grief and appreciation for her. Albert Einstein wrote a letter to the *The New York Times* in which he memorialized "Fräulein Noether [as] the most significant mathematical genius thus far produced since the higher education of women began."[340] Equally compelling is Hermann Weyl's less well known memorial, which spoke not just to Noether's mathematical acumen but to her character and persevering spirit: "Emmy Noether, her courage, her frankness, her unconcern about her own fate, her contradictory spirit, were in the midst of all the hatred and meanness, despair, and sorrow surrounding us, a moral solace."[341]

———————

Noether was only one of the refugee scientists that Bryn Mawr accepted through the end of the war. The college had participated in the Committee's efforts since its founding, and Marion Edwards Park, the president of the college from 1922–1942, was an early member. Park helped foster a commitment at Bryn Mawr to helping displaced scholars, particularly women who weren't welcome at male-dominated institutions.[342] In 1939, Bryn Mawr extended a lifeline to another woman mathematician, Hilda Geiringer (September 28, 1893–March 22, 1973), just when it seemed that she and her daughter Magda would be sent to a concentration camp.

By the time Geiringer arrived in the United States, she had been seeking refuge for six years, following her dismissal from University of Berlin (now Humboldt University) in 1933. Geiringer was the first woman lecturer at the university, and the first woman in Germany to obtain *habilitation* (permission to teach at a university) in applied mathematics. At the time, the University of Berlin was a hub for the burgeoning field of applied mathematics and home to the newly established Institute for Applied Mathematics and led by Richard von Mises, a leading applied mathematician and Geiringer's mentor. A member of the Institute, Geiringer was on the forefront of applied mathematics in Europe, making key contributions to the field in the areas of statistics, probability, and plasticity theory. Together with von Mises and other applied mathematicians, Geiringer co-developed the slip-line theory of plastic deformation, a set of simplification techniques to analyze the conditions in metal deformation, still used today in a wide variety of fields from safety engineering to plate tectonics in geology. Under the Civil Service Act, both Geiringer and von Mises were removed from the university.

Turkey became a safe haven for about 190 Jewish intellectuals fleeing Nazi Germany, Geiringer and her daughter, and von Mises among them.[343] In 1933, Turkey instituted the University Reform Law No. 2252, meant to transform the country's higher education to compete with other leading nations. Turkish president Mustafa Kemal Ataturk welcomed refugee intellectuals to bolster the new nation's educational system. The day after the reform law was passed the University of Istanbul was founded, where Geiringer joined the faculty as Professor of Mathematics the same year.

While working at the University of Istanbul, Geiringer regained some of the intellectual freedom she had lost in Berlin. Over the course of her five-year contract, she continued her research in plasticity and statistics but also broke into genetics when she began to apply probability statistics to Mendelian genetics. She published eighteen articles in English as well as a number in Turkish, a language she did not know before living in Turkey, and published a calculus textbook in Turkish.[344] When her contract was not renewed in 1939, von Mises refused to stay at the university without her, even though his contract was still in place. Adding to their troubles, Ataturk died in 1938, and they feared that the welcoming atmosphere for refugees the president had created was too fragile to survive his death.

Going back to Germany was not an option as conditions for Jews had only worsened in their absence. In 1938, the Nazis had staged a two-day spree of organized violence against German Jews called Kristallnacht (the Night of Broken Glass). Jewish businesses, homes, and synagogues across the Reich were vandalized and destroyed, and thirty-thousand Jewish men were rounded up and sent to prisons and concentration camps. This escalation of Nazi violence set off a mass exodus of Jews from the country. Resolving to flee to the United States, von Mises secured a position at Harvard, and with it, his non-quota visa. But Geiringer had no such luck finding a position in the United States. Without a job she could not secure a non-quota visa, and there were no available quota visas.

Geiringer took refuge with her brothers in London for a time, but when she and Magda were en route to the Mediterranean for a break, the war officially broke out. They were stopped in Lisbon, Portugal because their German passports were no longer valid for entry back into England. Stranded in Lisbon, Geiringer and Magda were threatened with deportation to a concentration camp. Meanwhile, in the United States, von Mises along with Albert Einstein and Oscar Veblen worked to find a position for Geiringer, homing in on Bryn Mawr and Smith College, another women's college. Both Geiringer and von Mises grew desperate. In his diary, von Mises describes himself as "disturbed," "worn out and almost desperate," and Geiringer even proposed that they marry for her visa.[345] Finally, Bryn Mawr came through with aid from the Emergency Committee, and Geiringer and Magda boarded a ship for New York.

←

HILDA GEIRINGER

Hilda Geiringer was the first woman
lecturer at the University of Berlin
and the first woman to obtain
habilitation in Applied Mathematics
in Germany. She made significant
contributions to slip-line theory and
Mendelian genetics.

→

TILLY
EDINGER

Tilly Edinger founded the subfield
of paleoneurology, the study
of fossilized brains.

Geiringer spent six years at Bryn Mawr before accepting a position at Wheaton College, a women's college in Massachusetts, as head of the Mathematics Department. But in all this time, she did not stop trying to find a position at a university. Bryn Mawr and Wheaton were mainly undergraduate institutions with little room for Geiringer to pursue her own research outside of teaching. In a 1953 letter to the president of Wheaton College, Geiringer described her desire for research:

"I must work scientifically...It is perhaps the deepest need in my life." Despite Geiringer's qualifications, she could never find another university position. In the words of a professor from Tufts University to Geiringer "...it is not merely prejudice against women, yet it is partly that, for we do not want to bring in more if we can get men."[346]

One woman who did manage to hold onto her position following the 1933 Civil Service Act in Germany was paleontologist Tilly Edinger (November 13, 1897–May 27, 1967). Unlike Noether and Geiringer, Edinger worked at a private institution when the law was passed, which shielded her for a time. She continued as Curator of Vertebrate Fossils at Senckenberg Museum of Natural History in Frankfurt for five years after the Civil Service Act was passed, but as a Jew, she still faced the possibility of anti-Semitic violence. She tried to make herself as invisible as possible by entering through the museum's side door and removing the nameplate from her office.[347] Edinger was also gradually becoming deaf from otosclerosis, a disease of the inner ear. Eventually she stopped attending professional meetings in which she had to sit in the front to hear because she feared drawing too much attention to herself. She faced heightened scrutiny for her deafness, a disability that was considered unacceptable under Nazi ideals of genetic purity. Under the 1933 Law for the Prevention of Offspring with Hereditary Diseases, thousands of deaf Jews were forcibly sterilized and hundreds were targeted and murdered over the course of the war.[348]

Despite the risks she faced, Edinger was reluctant to leave Germany; she loved her work at the Senckenberg where she had privileged access to fossils to study. Edinger had by this point become an internationally renowned paleontologist and had established the subfield of Paleoneurology, the study of fossil brains. Family friend Alice Hamilton later recalled in her book *Exploring the Dangerous Trades* that Edinger was resolute about staying in Germany, saying, "So long as they leave me alone I will stay. After all, Frankfurt is my home, my mother's family has been here since 1560, I was born in this house. And I promise you they will never get me into a concentration camp. I always carry with me a fatal dose of veronal."[349] But after Kristallnacht, Edinger knew staying in Frankfurt would be impossible. Once she decided to leave, paleontologists in the United States sprung to action to find her a position.

Edinger applied to the American consulate for immigration to the United States and was assigned quota number 13,814 in 1938, but her number was not expected to be called until the summer of 1940. With the help of the Emergency Association of German Scientists, Edinger was first able to find temporary safety in London, where she was considered an "enemy alien." She took with her only ten German marks, two spoons, two forks, and two knives.[350] For a year, she made a living translating medical texts from German to English, with additional support from family members in the city. Growing restless, she attempted to earn non-quota status, but since she did not hold a teaching position at "an institution of learning," her request was denied.[351] Fortunately, her quota number was called a little early, and she arrived in New York on May 11, 1940 aboard the *Britannic*.

With the help of the Emergency Committee, Edinger was able to obtain a research associate position at the Museum of Comparative Zoology at Harvard, a position she held until her retirement in 1964. In Germany, Edinger's last hope of someday returning to Frankfurt vanished when she received word that her aunt had died in 1943. In a letter to colleague Sir Arthur Smith Woodward, Edinger wrote, "My last tie with anybody in Germany ended with a terrible shock last year when I was informed...that my father's sister in Berlin, the woman I loved best in the world, has committed suicide when she was deported—at the age of 84½ years!"[352] Before that, her brother Fritz had died in a concentration camp. With no family left in Germany, Edinger decided to make the United States her permanent home, becoming a citizen in 1945.

Giants in their chosen fields, Neother, Geiringer, and Edinger were saved by their work. But for every woman who did find refuge in the United States and elsewhere, there were many more who did not, including Leonore Brecher, a pioneering Romanian biologist who died in an extermination camp in Minsk in 1942 after years of trying to escape. Those that were saved did not survive on their own; they were supported by the international cooperation of scientists and activists who came together and worked tirelessly to make a place for their fellow scientists. This era in science history serves as a reminder of what can be lost when paths to refuge are closed by fear and discrimination, but also what can be gained through collaboration and community.

CHAPTER 17

Nature's Housekeepers Begin a Movement

BEFORE CARSON

The 1962 publication of Rachel Carson's *Silent Spring* is often cited as the beginning of the environmental movement in the United States. Rachel Carson (May 27, 1907–April 14, 1964) was a marine biologist and conservationist, and her book *Silent Spring* documented and narratively illustrated the devastating effects that pesticides such as DDT had on various species and on ecosystems. In many ways, *Silent Spring* did signal a sea-change in environmental activism; it prompted widespread public concern about the unchecked use of pesticides and mobilized the federal government to legislate sweeping environmental protections. The impetus for the Clean Air and Water Acts and the founding of the Environmental Protection Agency in the United States have been credited to Carson's influence. But the story of women campaigning on behalf of nature and the environment begins much earlier. From bird and wildlife conservation to water rights, food protections, and urban sanitation, women have been a loud and astonishingly effective voice in the realm of conservation and ecology since the beginning of the twentieth century.

The early decades of the twentieth century witnessed radical reforms in both social and political policy in the United States, a period that has been termed the Progressive Era (see page 137). Progressive reformers tackled a range of social ills brought on by rampant industrialization and expanding urbanization. Concerns about industrially produced food, water contamination, waste, overcrowding, and the overall safety and sanitation of city living occupied the agendas of many Progressive reformers. Alarmed by the deteriorating

← Photographed in the woods near her home, Rachel Carson published *Silent Spring* in 1962. The book's publication has been widely credited as beginning the modern environmental movement.

conditions in their immediate environments, middle-class women in particular sought out ways to address the unregulated forces that threatened the welfare of their children and communities. Even though women were still excluded from the positions in local and federal government with the power to rein in industry's excesses, wives, mothers, and housekeepers argued that they had unique skills and experiences to offer Progressive causes. As Adella Hunt Logan of the Tuskegee Women's Club explained in 1912,

> *"Good women try always to do good housekeeping. Building inspectors, sanitary inspectors, and food inspectors owe their positions to politics. Who then is so well informed as to how these inspectors perform their duties as the women who live in inspected districts and in inspected houses, who buy food from inspected markets?"*[353]

Logan's Tuskegee Women's Club was just one of hundreds of women's organizations around the country with environmental issues on their roster of reforms. Still mired in Victorian social norms from the nineteenth century, middle-class American society relegated women to the domestic sphere, but women reformers turned this into an unlikely advantage. In a 1915 handbill, Susan Fitzgerald of the Woman Suffrage Party wrote, "We are forever being told that the place for women is in the HOME." But if the home was conceived of as bigger, much bigger than the four walls of their own houses, they could extend their reach and influence into their cities, their states, and, ultimately, their country. "Women are by nature and training, housekeepers," Fitzgerald continued, "Let them have a hand in the city's housekeeping, even if they introduce an occasional house-cleaning."[354] As housekeepers, they were tasked with taking care of all these spaces. As mothers, they were compelled to make them liveable spaces that nurtured the next generation. Their argument for authority in such matters was rooted in essentialist beliefs about the supposed nature of womanhood, which made them more suited than men to clean up and preserve nature's "house." In a 1909 address to the Conservation Congress, Mrs. Overton Ellis of the General Federation of Women's Clubs explained this uniting principle of the women's clubs, saying, "Conservation in its material and ethical sense is the basic principle in the life of woman[...] Woman's supreme function as mother of the race gives her a special claim to protection not so much individually as for unborn generations."[355]

Although not a mother herself, Ellen Swallow Richards in many ways exemplified the principles of womanhood that rallied the women's clubs.

←

ELLEN SWALLOW RICHARDS

Ellen Swallow Richards was the first female student at the Massachusetts Institute of Technology and was a leader in advocating for food and water purity laws.

→

MARY CHURCH TERRELL

Mary Church Terrell, the first president of the National Association of Colored Women, advocated for better housing conditions and sanitation services for Black people and neighborhoods.

The first female student at the Massachusetts Institute of Technology (MIT) during the 1870s, Richards called on women to harness their knowledge of daily life—cooking, cleaning, and economic resourcefulness—and combine it with the science of chemistry to meet the challenges of life in an industrial society. At MIT, Richards went from chemistry student to teacher, starting an all-women laboratory at the university. Richards taught her students nutritional science, domestic application of modern technology, and scientific preparation of food. While Richards was bringing the science of the home into the all-male halls of MIT, domestic science was taking root in other parts of the country with the first official college course, Chemistry as Applied to Domestic Economy, at Iowa State College in 1871.[356] Women's knowledge in the home was being taken seriously and gaining legitimacy, creating a base of authority for Progressive women reformers at the turn of the century.

While the science of healthy food and sanitation began in the home, Richards argued that it could not be isolated there. A healthy home was the fundamental starting point for creating a healthy community and environment. In 1892, Richards brought a new way of thinking to questions of the relationship between nature and the built environment to the United States when she introduced Americans to Ernst Haeckel's concept of *oekologie*, what we now call ecology. For Richards, ecology was not merely biological, but encompassed a complex network of interactions between humans and nature and between the home, the economic, and the industrial.[357] When industry disrupted the delicate system, an educated community of women had the power to bring balance back into the system. "It is for women to institute reform," she told a women's club in Poughkeepsie, New York in 1879. Further, she argued, "It is not an easy task we have before us. So long as we are content with ignorance, so long we shall have ignorance; but when we demand knowledge, because we know the value of knowledge, then we shall succeed."

Not content with giving speeches, Richards put her theory of ecology and domestic science to the test in Massachusetts. In 1890, she and her friend Mary Abel opened the New England Kitchen, a low-cost, public kitchen that implemented nutritional science in food preparation for working-class and immigrant families in Boston. At her women's lab at MIT, she and her students undertook a massive study of the state's food supply, finding that industrially produced food was contaminated with chloride, mahogany, and other dangerous chemicals. These results led to the state's passage of the first food purity laws in the United States, predating the Food and Drug Act of 1906 by more than twenty years. In 1886, Richards began the most comprehensive water-quality survey in the

United States to date. Working around the clock for two years, Richards analyzed 40,000 water samples collected from the water supply, and sewage from eighty-three percent of the population.[358] An outgrowth of her study was the first water pollution map in the country, which visualized traces of chlorine from salt used in food and industry in water. Based on Richards's research, Massachusetts opened the new State Water Laboratory to monitor water pollution in the state. Starting with women's basic housekeeping concerns of food and sanitation, Richards built a movement that extended this domestic environmental consciousness to her community, to her city, and to her state.

EXPANDING THE MOVEMENT

As effective and influential Richards was, her reforms did not resonate with all women, particularly those in immigrant communities and among the working class. The careful scientific preparation of her food in the New England Kitchen was a poor substitute for immigrants who preferred foods from their home countries. Many resented what they saw as an attempt to impose American values on immigrant communities.[359] Richards also took a disparaging view of non-Western people, claiming in her book *The Art of Right Living* that "native religions" stood in the way of true reform in other countries.[360] The disdain Richards felt for the experiences and cultures of immigrants and non-white people was pervasive throughout white women's clubs of the Progressive Era. Despite its wave of social reforms, this was still the era of racial segregation. The General Federation of Women's Clubs, for instance, refused to give credentials to Black women at its conventions.[361] Black women formed their own clubs, like Adella Hunt Logan's Tuskegee Women's Club and the National Association of Colored Women (NACW) led by Mary Church Terrell.

Women's clubs run by Black women were better suited to meeting the needs of the working class as they shared more with the communities they wanted to serve than white women like Richards and the women of the GFWC.[362] The poor environmental conditions in urban areas and rural communities were even worse for Black people, who experienced higher rates of infant mortality and poverty while also having to fight racial oppression. Rebecca Cole, the second Black woman to receive a medical degree in the United States, argued that Black people had no choice but to live in unsanitary conditions due to discriminatory housing practices and exploitative landlords who collected inflated rent on overcrowded and unclean buildings. In a 1896 article in *The Woman's Era*, Cole showed how these conditions compromised the health of Black people and, in turn, justified stereotypes about Black people being inherently weak and prone to disease. She insisted that "We must teach these people the laws of health," and "we must preach this new gospel, that the respectability of a household ought to be measured by the condition of the cellar..."[363] For Cole as it was with Richards, the home was the starting place for much larger reforms.

Mary Church Terrell and members of other Black women's clubs harnessed the language of housekeeping much like white women's clubs, but focused on the conditions for Black people in particular.[364] Just two years after Cole's article, Terrell also targeted housing in a pamphlet titled "The Progress of Colored Women," writing,

> *"Believing that it is only through the home that a people can become really good and truly great, the National Association of Colored Women has entered that sacred domain. Homes, more homes, better homes, purer homes is the text upon which our sermons have been and will be preached. [...] By the Tuskegee club and many others all over the country, object lessons are given in the best way to sweep, dust, cook, wash and iron."[365]*

The role of motherhood figured prominently in Terrell's call for organizing because Black children were the "future representatives of the race" and that "nothing lies nearer the heart of the National Association than the children..."[366] Beyond individual homes, the NACW sponsored "cleanup days" and called on local governments to direct sanitation services to their neighborhoods.[367]

BEYOND THE CITY

The ideology of women's housekeeping also encompassed forests, wildlife, and nationally treasured landscapes. Mrs. Robert Burdette as president of the California Federation of Women's Clubs maintained that "[t]he preservation of the forests of this state is a matter that should appeal to women," for the preservation of the forests would ensure the health of California's people.[368] From coast to coast, women's groups were working to save the country's natural wonders and historic sites "from utter destruction by men," she wrote. On the eastern coast of the United States, the Florida Federation of Women's Clubs (FFWC) was undertaking the first effort to save the ecologically crucial tropical wetlands of the Everglades.[369] Founded in 1895, the FFWC tackled urban sanitation and public health on one front, the conservation of forests and birds on another. The FFWC succeeded in establishing Florida's first state park, Royal Palm State Park, in 1916.

↑

MARJORY STONEMAN DOUGLAS

Beginning her career as a journalist, Marjory Stoneman
Douglas became the leading voice in saving the
Everglades of Florida with her 1947 book
The Everglades: River of Grass.

Just the year before the establishment of Royal Palm, Marjory Stoneman Douglas (April 7, 1890–May 14, 1998) moved to Miami, Florida where she began working with the FFWC. A gifted writer and reporter, Douglas used her platform as a daily columnist at the *Miami Herald* to highlight and advocate for the social and environmental causes important to FFWC. Originally from Minneapolis, Minnesota, Southern Florida became for Douglas the locus of her activism and the setting for much of her fiction and non-fiction writing. She said that it was in her column, "The Galley," "that I started to talk about Florida as a landscape and as geography, to investigate it and to explore it."[370]

She left her staff writing job at the *Miami Herald* in 1923, publishing more ardent and overt defenses of non-human nature and its conservation as freelance writer. Even her works of fiction featured nature in some way. The Everglades was particularly cherished place for Douglas, who saw it as both "the [economic] wealth of south Florida" and "the meaning and significance of south Florida."[371] For decades, the Everglades was an endangered natural landscape, threatened by corporate agriculture and encroaching real estate development. In 1929, the Tropic Everglades National Park Association asked Douglas to join them in their effort to turn the Everglades into a national park.

Douglas's study of the Everglades culminated in her 1947 book *The Everglades: River of Grass*, which opens with the iconic line: "There are no other Everglades in the world."[372] The book contained five years of extensive research into the ecology of the Everglades, which was still widely unknown at the time. Her book was at once a scientific exploration of the ecosystem of the Everglades and an artfully composed homage to the singular beauty and complexity of the tropical wetlands.

The Everglades: River of Grass sold out in its first month and has been compared to Carson's *Silent Spring* in the way it called people to action to protect nature. The year of the book's publication, Douglas had the pleasure of seeing Everglades National Park formally dedicated by President Truman.[373] Douglas's advocacy for the Everglades and southern Florida would continue for the rest of her life. In 1969, she founded the conservation group Friends of the Everglades, and she was frequently involved in local politics in an effort to oppose draining Florida's wetlands for development that threatened the state's natural landscapes.

A NEW ERA FOR THE MOVEMENT

From the local level of communities, to the regional level of south Florida, to the national level of federal legislation, women were fundamental agents of environmental change in the United States in the early twentieth century. Rachel Carson's forebears demonstrated what was possible when the experiences and knowledge of women was taken seriously. However, the momentum of the Progressive Era's brand of nature's housekeeping eventually tapered off, and emerging feminist movements sought to discard the ideology of separate spheres that had powered it. More than that, men began to push back against reformers' attacks on industry by placing care for nature in opposition to science. Psychologist and educator G. Stanley Hall claimed that "caring for nature was female sentiment, not sound science," while others likened the billowing smoke of industry with masculinity and dismissed those opposed as feminine and sentimental.[374] The authority that women had wielded in environmental matters was eclipsed by a decidedly more masculine technical and economic approach, and many women were forced to resign from leadership positions in environmental organizations.[375]

When Carson arrived on the scene in the early 1960s, she was subject to such attacks on her expertise. Much like Douglas's *The Everglades: River of Grass*, Carson's *Silent Spring* was much more like literary nature writing than a scientific text. That she did not have an institutional affiliation at the time only further proved to her detractors that she was unqualified to indict the chemical and agricultural industries. Derided with thinly veiled misogynistic stereotypes as a hysterical or emotional spinster, Carson was often seen as little more than an amateur. Despite this criticism, many scientists found her work to be sound, and it seemed that the federal government and the American public took the same view. Carson died in 1964, but she had already set in motion new ways of thinking about the environment. In 1967, the Environmental Defense Fund was established and was instrumental in eliminating the use of DDT, followed by the Nixon Administration's formation of the Environmental Protection Agency in 1970.

Since Carson, environmental movements led by women have sprung up around the world. But this earlier history of women as nature's housekeepers shows that women have been advocating for safer environments, whether that be in cities or forests, for much longer and in many different ways than our current environmental movements. Women speaking for nature, advocating for safe and healthy environments from the home to the fragile natural landscapes in which we live is a long tradition, unbroken in our own time.

Supreme Court of the United States

No. 1 ———— , *October Term, 19* 54

Oliver Brown, Mrs. Richard Lawton, Mrs. Sadie Emmanuel et al.,

Appellants,

vs.

Board of Education of Topeka, Shawnee County, Kansas, et al.

Appeal from *the United States District Court for the* ————————————————
District of Kansas.

This cause *came on to be heard on the transcript of the record from the United States*

District Court for the ———————— *District of* Kansas, ————————————
and was argued by counsel.

On consideration whereof, *It is ordered and adjudged by this Court that the judgment*
of the said District ———————————— *Court in this cause be, and the same is*
hereby, reversed with costs; and that this cause be, and the same
is hereby, remanded to the said District Court to take such
proceedings and enter such orders and decrees consistent with
the opinions of this Court as are necessary and proper to admit
to public schools on a racially nondiscriminatory basis with all
deliberate speed the parties to this case.

Per Mr. Chief Justice Warren,

May 31, 1955.

CHAPTER 18

The Double Bind in the Sciences

In the spring of 1976, the director of the Northside Center for Child Development in New York, psychologist Mamie Phipps Clark, was giving a tour and interview at the center she had founded thirty years before. Ed Edwin, the interviewer, asked Clark basic questions about the purpose of the center, which provided psychiatric therapy and social work for children. Then, after Clark described growing up as a Black girl in a small segregated town in Arkansas, Edwin asked, "Do you recall when you heard about the first lynching?" Clark, then fifty-nine, told the story of the lynching of a Little Rock prison inmate around the time that she was six years old. Asked whether this experience lingered, Clark replied "I can sense the emotion of it now. You know, I can sense the fright of it—now."[376]

In the United States, the legacy of chattel slavery and the institutionalization of racism under Jim Crow in the second half of the twentieth century are the most important factors conditioning the experience of Black women scientists. The lived experience of segregation and lynching loomed large in both the personal and scientific lives of women like Clark who carried these memories with them into their research. For Black women, already marginalized in science by their gender, access to sufficient educational opportunities to train as scientists, engineers, and doctors is a central factor in their underrepresentation in science today, and in the past.

Clark attended a segregated school for all of her primary education, and in the 1976 interview, she remembered passing the white students, whose school was on the opposite side of town, every morning as they walked to class.[377] In 1934, Clark earned a scholarship to Howard University, a historically Black university in

← *Brown v. Board of Education*, a landmark 1954 civil rights case in the US supreme court which ruled school segregation unconstitutional.

Washington D.C., to study mathematics and physics. Historically Black colleges and universities (HBCUs) like Howard were essential institutions that provided access to science education at a time when major research institutions were still segregated. Howard provided students like Clark an opportunity to receive an education equal to that of their white peers.

At Howard, Clark met her future husband, psychology student Kenneth Bancroft Clark. He persuaded her to change her major to psychology. After graduating *magna cum laude* in psychology in 1938, she spent the summer at Charles Houston's law office, which was engaged in work on a number of important Civil Rights cases leading up to the Supreme Court ruling in the *Brown v. Board of Education* case that school segregation was unconstitutional.[378] Exposure to lawyers including Thurgood Marshall who were preparing segregation cases was what Clark called "the most marvelous learning experience."[379] She said she gained a "sense of urgency, you know, of breaking down the segregation, and the whole sense of really, blasphemy, to Blacks…"[380] Clark's experience at the law firm influenced her graduate work at Columbia University.

Clark and Kenneth were collaborating at the same time she was completing her PhD dissertation on development and mental ability in children. Their work together focused on racial identification among children, demonstrating that children learn to identify their own race and that of others as early as seven years old and show sensitivity to racist perceptions of themselves and their peers. Their research found that children who attended segregated schools were more likely to accept racial injustice as a fact of life, whereas students from integrated schools were more conscious of inequity.[381] This is something Clark herself experienced as a child, noting that while her father was well-respected and that gave her family certain privileges, the day-to-day experience of segregation was routine. "For instance," she said "when we went to football games, which were out of town, you know, that we would have to take our own lunch, or that we would have to find our own bathroom facilities."[382]

Clark's most famous research was the "Doll Test" in which Black children were presented with identical dolls with different colored skin and asked a series of questions about which dolls they preferred. Directions such as, "Give me the doll you like to play with or like best," were designed to assess racial preference, and questions such as "Give me the doll that looks like you," measured racial self-identification.[383] The Clarks found that majority of Black children preferred dolls with light skin and hair while discarding the doll with dark skin and hair and assigning negative attributes to them. The Doll Tests, published in 1947, were incorporated into expert testimony by social scientists in Civil Rights cases and a Supreme Court Brief by the National Association for the Advancement of Colored People in the *Brown v. Board of Education* Decision in 1954.[384]

Clark's life and work was deeply intertwined with the Civil Rights movement, so much so that her faculty advisor at Columbia, Henry E. Garrett, would later argue against school integration in churches, attempting to counter the legal challenge supported by Clark's own research. After she received her doctorate, Clark had difficulty finding work to match her high qualifications. She, instead, created her own opportunities with the Northside Center for Child Development in New York, which she founded in 1946. At the time, it was the only organization in Harlem providing mental health services to Black children. Clark's experience of higher education was in one sense atypical—she was the first Black woman PhD at Columbia and the first Black woman in the country to receive a PhD in psychology. But in many ways, hers was the typical experience of Black women scientists, who in the mid-twentieth century United States, and still today, find themselves subject to the "double bind" of being Black and a woman in the sciences.

RACE AND GENDER: THE DOUBLE BIND

In 1975 the American Association for the Advancement of Science (AAAS) convened a small conference by and for marginalized women in science called "The Double Bind: The Price of Being a Minority Woman in Science." The goal was to gather information about the experiences of working Black and Brown women scientists and draft recommendations for ways to improve education, professional opportunities, and policy. "Our mission at this meeting was clear," the organizers wrote in the conference proceedings. "We wanted to find out how and why we had made it and others had been left behind; how our sisters had handled personal and societal problems for childhood until the present."[385]

One of the conference's major findings was that Black and Brown women (called "minority women" in the report) had largely been left behind by organizing efforts for women in the sciences and the larger Women's Movement of the late 1960s and 1970s because these efforts were largely focused on the experience of "majority" white women.[386] The conference participants answered questions about their childhoods and early education, college and graduate school, and their professional careers, paying special attention to the ways that their race or ethnicity intersected with their gender in these environments and how their interest in science further shaped their lives. Most of the older Black women at the conference reported attending segregated schools as children, which "usually suffered from inadequate facilities, from the building to the books, although some students were not always aware of the difference unless or until they had exposure to other schools." Even in integrated schools, teachers often had lower expectations of Black and Brown students.

Many women reported that their interest in science further marked them out as different from their peers at a time when science was not considered an appropriate or reasonable career interest for girls. The conference report said that this contributed to a sense of isolation that many women at the conference had felt throughout their lives in science: "[B]y the time they finished high school these young women's interest in science had already established a pattern of differentness in their lives."[387] The conferees discussed the way this differentness was a product of the interconnected oppressions of being non-white and women. As the report noted, "Loneliness, pressure to choose a traditional career, to marry, to remain in or return to the community of their youth—in summary, to fill cultural role expectations—was constant..."[388]

The experience of being a woman of color in higher education took on a different character for the women who were educated at HBCUs. As the Double Bind conference reported in 1975, Black colleges have been the main producers of Black scientists in America.[389] As of 2004, HBCUs made up only about one

For Black women, already marginalized in science by their gender, access to sufficient educational opportunities to train as scientists, engineers, and doctors is a central factor in their underrepresentation in science today, and in the past.

percent of all colleges and universities in the United States, but between 1994 and 2001, they accounted for nearly thirty percent of all science and engineering degrees awarded to Black people.[390] Historian of science Olivia Scriven has documented the importance of HBCUs to the production of Black scientists, beginning with the founding of such institutions in the nineteenth century to provide education to Black people who were barred from attending mainstream universities. That HBCUs produced more Black scientists than other institutions is the result of a complex interaction of social and cultural factors that shaped the way Black people have been educated in the United States, often according to racist precepts about the "proper" place of Black people in a white supremacist society.

Scriven traces the way that science education for Black people has been systematically undervalued in a country that prizes science and engineering degrees and careers for white students, even at moments in which the production of scientists was linked to the national interest.[391] In post-World War II United States, people like engineer and science administrator Vannevar Bush pressed for increased federal investment in science and engineering education as a matter of national priority. The resulting influx of money allocated by the National Defense Education Act of 1958 was spent at institutions with substantial research capability, something that chronic underfunding had prevented HBCUs from developing. In this respect too, women's colleges benefitted much more from this new source of funding due to their longer, better-funded history of developing research programs.[392] Excluded from the white college and universities, including many women's colleges, Black women who wanted to pursue science largely did so at HBCUs.

When the Black colleges were established after the Civil War, there was significant debate about the type of education they should offer. Concerned with keeping Black people "in their natural sphere," white segregationists argued that Black people should be given practical education that would prepare them for service work.[393] Black leaders also debated the type of education that was most useful for Black people, but this debate was about whether science and engineering, or a more classical liberal arts education, would better prepare Black people to *avoid* going into service and industrial work.[394] Another compounding factor was that many HBCUs were actually founded and run by white people, often as arms of missionary and religious organizations. Spelman College in Georgia was one such institution, founded by two white Baptist missionaries as a seminary for formerly enslaved Black girls.[395]

RESHAPING SCIENCE EDUCATION IN HBCUs

Flemmie Pansy Kittrell (December 25, 1904–October 3, 1980) was born in North Carolina to a family of sharecroppers, and as a young woman, she worked as a domestic servant. She received a bachelor's degree at the Hampton Institute in Virginia, which, like Spelman, was a private school founded by missionaries after the Civil War to educate freed Black people.

As part of her scholarship to Hampton, Kittrell did domestic work at the institute to cover her tuition.[396] After graduating, Kittrell went to Bennett College in North Carolina, where she served as Dean of Students and Director of Home Economics. At Bennett, she was encouraged to pursue graduate education, eventually enrolling at Cornell, where she received a master's degree in Home Economics in 1930, followed by a PhD in the same field five years later.[397] She was the first African American woman to earn a doctoral degree in Home Economics.[398]

Like Clark, Kittrell's interest in child development grew during her graduate work. Her master's thesis and doctoral dissertation investigated domestic life in her home state of North Caroline, the latter focusing on infant feeding practices in a small community.[399] After Kittrell's father died in 1919 when she was five years old, she observed her mother's running of their household and learned from her that "[c]hildren were learning, apparently all the time." She cited this experience growing up as formative for her interest in family dynamics and home economics.[400] Kittrell was trained in a tradition of home economics that emphasized the importance of the family as the basic unit of the nation and that the perfection of domestic life had far-reaching implications.[401]

At Bennett, the home economics department was very new on her arrival in 1938, consisting of only one teacher and a poorly equipped lab. In two years, Kittrell expanded the department to include more instructors and better facilities, including a practice home where students could get hands-on experience in the principles of domestic engineering and home economics.[402] In 1940, Kittrell returned to Hampton while it was playing host to military trainees and and she took on "war work," teaching the public about food rationing.[403] In 1944, she began working at Howard, which would be her home for the next three decades.

Kittrell served as the head of the home economics department at Howard, where she substantially reshaped the program and facilities, incorporating child development into the curriculum and created a preschool to serve as a kind of laboratory for her work on nutrition and development.[404] Nurse trainees also rotated through the department to get experience working with children and studying diet and nutrition.[405]

Kittrell was also heavily involved in outreach work abroad in the 1950s, traveling to India and various parts of Africa to teach about home economics and set up college programs. Some of this work was sponsored by the State

Department as part of its ideological imperative to demonstrate the superiority of American education and science. Kittrell and other Black academics were, as historian Allison Beth Horrocks agues, "[c]ommissioned to do this work because she could be presented as an exemplar who had thrived within the (segregated) US education system."[406]

The 1975 AAAS conference report marked an important moment in the timeline of Black women in science in the United States, giving for the first time a sanctioned, institutional voice to the struggle that they had endured in a society dominated by the legacy of slavery and white supremacy. The conference proceedings offered an important critique of majority-white feminist movements within and without science, which had largely left Black women scientists behind. But even in a forum designed explicitly to speak to the experiences of marginalized women, many still struggled to articulate to conference organizers the role of racism in their lives and careers, simply because it was the ever-present background of their existence. "Racism and its manifestations and effects," the report read, "have been so intrinsically part of their lives that these women gave little time to the discussion of its manifestations, except in the context of particular experiences." Mamie Phipps Clark's 1967 reflection on the lynching that shook her community when she was a child bear out the observations of the Double Bind participants—white supremacy was not an event in the lives of these women so much as the fabric of the world they inhabited. That so few Black women had achieved stable careers in science by the 1970s, a full decade after the passage of the Civil Rights Act, was not an accident of fate or a result of a lack of interest. Instead, they faced structural oppressions and obstacles that were intrinsic to American society, and the history of science, in the twentieth century and which persist in our present.

Mamie Phipps Clark

(April 18, 1917–August 11, 1983)

Mamie Phipps Clark was an American social psychologist, who specialized in child development in Black children. Born in Arkansas, Clark drew on her early experiences as a Black child in the segregated American South to help children growing up with the same inequalities.

Clark started at Howard University in 1934, first majoring in mathematics and minoring in physics, but she switched to psychology after meeting psychology student Kenneth Clark, who would become Clark's husband and long-term professional collaborator. Clark graduated *magnum cum laude* in psychology before pursuing graduate studies. Her master's thesis, "The Development of Consciousness in Negro Pre-School Children," investigated the age at which young Black children become aware of their race, concluding that boys as young as three and four showed distinct racial awareness.

Clark went on to complete a PhD in psychology from Columbia University in 1943. She and Kenneth, now her husband, were the first two Black people to earn PhDs at Columbia. She also received a Julius Rosenwald Fellowship, and with their funding and her collaboration with Kenneth, she initiated the famous Doll Test, which showed that Black children in segregated schools were more likely to prefer dolls with white complexions and yellow hair while discarding the brown dolls with black hair and assigning negative traits to them. The study showed the devastating effects of school segregation on Black children.

Based on their research, Clark and Kenneth, testified in many school segregation cases in the South, and ultimately, Kenneth used their research to argue for school integration in the 1954 Supreme Court Case *Brown v. Board of Education*. This was the first time that social science was used in a Supreme Court case.

In 1946, Clark and Kenneth opened the Northside Center for Child Development, the only mental health organization for Black children in New York, and although Clark retired in 1976, the center is still open today. Clark was awarded the American Association of University achievement award in 1973, and ten years later the National Coalition of 100 Black Women awarded her the Candace Award for humanitarianism. Clark died of lung cancer in 1983.

CHAPTER 19

More than Astronauts

In the late spring of 1963, Soviet cosmonaut Valentina Tereshkova (born March 6, 1937) became the first woman to fly in space. Tereshkova grew up in Maslennikovo on a collective farm, about a hundred and sixty miles (250 km) northeast of Moscow near the Volga river.[407] As a young woman, Tereshkova was an avid amateur parachutist and a member of her local Young Communist League.[408] Tereshkova volunteered herself for the cosmonaut program after the Vostok 2 flight of Gherman Titov in 1961.[409] After an interview and medical testing, Tereshkova was invited to train with four other women cosmonaut candidates at the Cosmonaut Training Center in Zelenyy. Tereshkova was selected for the program in part because of her parachuting experience—the Vostok spacecraft used by the Soviet space programs parachuted back to earth, with the cosmonaut inside ejecting before impact and parachuting separately to the surface.[410] Tereshkova successfully piloted Vostok 6 through a three-day mission beginning June 16, 1963, safely returning to instant acclaim in Russia.

The Soviet press and space program celebrated Tereshkova's flight as tangible proof of the high status of women in Soviet society.[411] Yet, it was not purely progressive beliefs about women that propelled Tereshkova to space; the decision to train a woman for spaceflight in the USSR was made for largely political reasons. The Air Force General who oversaw the cosmonaut program first broached the idea in 1961, writing in his personal notes that women would eventually fly in space and should be trained sooner than later. "Under no circumstances," he continued, "should an American become the first woman in space—this would be an insult to the patriotic feelings of Soviet women."[412]

← Space Shuttle *Discovery* launching from the Kennedy Space Center, Florida in 1988.

In the early 1960s, the Cold War contest between the United States and the Soviet Union had reached a fever pitch: in 1962, the United States discovered an emplacement of Soviet nuclear missiles in Cuba, a mere ninety miles (about 150 km) from the American coast. But the existential threat of nuclear war was only one part of the standoff. The ideological components of the Cold War had far-reaching consequences for the history of science and technology, particularly in the field of spaceflight. The so-called 'space race' was kicked off in 1957 by the Soviet launch of the first artificial satellite, Sputnik I. For the USSR, spaceflight was a way to demonstrate the ideological superiority of socialism as it translated into technological innovation. Not to be outdone and risk the world perceiving their system of government and capitalist economy as weak, the United States quickly spooled up a space program that could match the USSR in heroic feats of technological spectacle.

Tereshkova's flight was part of this political spectacle. More women cosmonauts and astronauts would follow her into space, not just American astronauts including Sally Ride (May 26. 1951–July 23, 2012) but women from the European, Japanese, Indian, and Chinese space programs, all of whom grew up in the wake of the conflict between the USSR and the United States. But women astronauts and cosmonauts are only a handful of the women who have made their way through their country's space programs, although they have been the most visible. From clerical workers to food scientists, tens of thousands more women have made spaceflight their life's work.

WOMEN OUT OF PLACE IN SPACE WORK

In June of 1963, *Spaceport News*, the employee newspaper for NASA's Florida launch facility the Kennedy Space Center (KSC), published a special edition featuring the contributions of women to the American space program twenty years before Sally Ride made her historic spaceflight.[413] The paper's editors prefaced the issue by noting that while many of the jobs women did at KSC were not nearly as exciting as that of the Russian cosmonaut Valentina Tereshkova, who had become the first woman to fly in space only four days earlier, there was much to admire about the women who worked at NASA. In "[s]uch a highly technological and specialized field as space," they wrote, "[a] secretary, a file clerk, a typist, although performing relatively mundane duties, is by the nature of carrying out these duties relieving her boss so he (or she) may concentrate on more important matters."[414]

This flippant attitude toward clerical workers in the early days of American spaceflight was typical of a project dominated by engineers, military men, and bureaucrats. Women who were not themselves trained scientists or engineers were thought of as mere accessories to the men whose memos they typed and whose phones they answered. As the case of "human computers" demonstrates, decisions about what counts as "technical" work were often arbitrary and had much more to do with the gendering of certain kinds of labor than the actual technical content of that labor.[415] Making sharp distinctions between technical and nontechnical labor is one of the ways women have been consistently marginalized in the history of science and in the history of spaceflight.

In the same issue of *Spaceport News*, the editors of the paper reprinted a story from a local newspaper, the *Capeside Inquirer*, which featured interviews with some of the clerical workers at KSC. In asking these women "how they felt members of their sex could best aid space programs," it did not seem to occur to the reporter that women who worked for the space program contributed to space programs the same way men did, by *doing their jobs*. Tens of thousands of women worked for NASA and its contractors in the 1960s, but where their work was classified as clerical, "pink collar" labor, their stories have been forgotten.[416]

WOMEN SCIENTISTS AND ENGINEERS AT NASA

Beyond the thousands of women in clerical positions within the space program, even those with "technical" jobs have often been overshadowed by the few women who became astronauts. Astronauts themselves were thought of by mission planners and engineers as a component system in the technological complex of spaceflight, one which required careful maintenance. Rita Rapp (June 25, 1928–July 12, 1989) was an Ohio-born physiologist who specialized in the biomedical aspects of high g-force flight. Working at the aeromedical labs at Wright-Patterson Air Force Base after graduating from the St. Louis Graduate School of Medicine in 1953, Rapp studied the effects of g-forces on the circulatory systems of the body.[417]

Rapp joined NASA in the early 1960s to work on problems related to the effects of spaceflight on the body and develop experiments for astronauts to conduct while in orbit during the Gemini program. She is most well-known for her Apollo work as part of the Food Systems Team, for which she devised stowage solutions for the astronauts' meals and transformed the early space food from something that had to be squeezed out of a tube into something that could be eaten with utensils.[418] Rapp explained to a journalist that her job was "viewing food as the hardware—it's my job to see it's on board the spacecraft."[419]

↑

VALENTINA TERESHKOVA

The first woman to fly in space, Valentina Tereshkova became a hero in the Soviet Union after spending nearly three days in orbit aboard Vostok 6.

→

RITA RAPP

Physiologist Rita Rapp worked at NASA in the 1960s developing food stowage systems for the *Apollo* astronauts.

Throughout her long career at NASA, Rapp authored dozens of papers on space medicine before her death in 1989.[420]

The maintenance of the "human systems" components of spaceflight was a challenge that necessitated the creation of a new medical subspecialty, aerospace medicine. In 1961, Dee O'Hara (born August 9, 1935) became part of NASA's aerospace medical team when she joined Project Mercury as staff nurse. Born in Idaho in 1935, O'Hara trained as a nurse in Oregon before joining the Air Force in 1959 as a Lieutenant.[421] O'Hara was stationed at Patrick Air Force Base at Cape Canaveral, Florida, just south of the NASA and Air Force launch facilities that would become KSC. O'Hara was transferred to a NASA assignment in 1960 to help set up the agency's aeromedical laboratory for the nascent human spaceflight programs.[422] O'Hara became the first astronaut nurse, working alongside the astronauts' physician to monitor their physical condition and see to their specialized medical testing and preparations for spaceflight.

O'Hara remembered that her position as the only nurse at KSC, and as one of the few women in general, was not isolating for her personally. "So at the Cape," she said in 2002, "it really was totally a man's world. I was the only female in Hangar S, except for maybe there was probably a secretary here or there. It was all men, and never once was I ever discriminated against or never made to feel uncomfortable." She admitted that she "wasn't really the sensitive kind, if you will, or looking for every little nuance," but has maintained that her relationships with the astronauts, engineers, and administrators at NASA were all positive.[423] But O'Hara remained one of the few women involved with early American spaceflight for most of her career. And she never was never a nurse to a woman astronaut.

AMERICAN WOMEN FLY IN SPACE

When American women were finally admitted to the astronaut corps, it was part of the eighth group of candidates selected by NASA in 1978, nineteen years after the first seven men were selected to fly in space.[424] This group was divided into pilots and mission specialists, reflecting the different roles astronauts would serve in operating the new Space Shuttle. The women selectees in Group 8 were all mission specialists: Anna Fisher, Shannon Lucid, Judith Resnik, Sally Ride (see page 216), Rhea Seddon, and Kathryn Sullivan. These six women, and the three non-white men in the group, were dubbed by Sullivan as the "nine strange people" among the otherwise familiarly white, male astronaut corps.[425]

When NASA introduced women to the astronaut corps in the late 1970s, many new cultural, technological, and political shifts followed. As a quasi-military organization with a massive bureaucracy and a core competency designed around an all-male "human system" for spaceflight, making the changes required was at times labored. For the women astronauts, the media attention generated by their selection was itself an obstacle in their early training, as they were beset by reporters even in places where they were supposed to be off limits.[426] On such field trips away from Johnson Space Center (JSC), where most of their training took place, NASA took great care to ensure privacy for each astronaut, not only as a matter of comfort but also to avoid any suggestion of sexual impropriety among a newly mixed gender astronaut corps.[427] Historian Amy Foster points out that at NASA the majority of engineers and scientists had only ever worked with women who were secretaries and clerical workers; the introduction of women with equal status or celebrity was a difficult transition for some.[428]

Even the physical spaces of spaceflight had to be transformed to accommodate women at NASA. At JSC, where astronauts trained both their minds and bodies, there were no women's locker rooms in the gymnasium and no sports bras in the inventory of athletic wear provided for astronauts.[429] Carolyn Huntoon, a researcher who was already a fixture at NASA when Group 8 was selected, was the first woman to serve on the astronaut selection committee, and she oversaw many of the changes to facilities and protocol to integrate women. Advocating on behalf of the women astronauts, often quietly in the background, Huntoon saw to it that *attitudes* toward women astronauts were also adjusted, always on guard for double standards about everything from performance expectations to dress code.[430]

The integration of women astronauts also highlighted the ways that spaceflight technology had been designed around the assumption of male astronauts. The contractor that made space suits for NASA in the late 1970s redesigned the EVA (extravehicular activity) suits to better fit women astronauts, making them smaller. But because of the rigid design of the upper part of the suit and the tendency for women astronauts to have narrower shoulders and less upper body strength than men, the suits were still too cumbersome for many women to perform EVA tasks, effectively excluding them from participating in spacewalks.[431] In fact, the first spacewalk to consist of only women astronauts did not happen until October of 2019, fifty-four years after Soviet cosmonaut Alexei Leonov became the first man to walk in space.

Because of their public profiles, astronauts are often some of the most well-known scientists, although their scientific accomplishments are usually overshadowed by spaceflight milestones. Mae Carol Jemison (born October 17, 1956), for example, was the first African American woman to fly in space. But before she joined NASA, she was already an engineer and a working physician. Born in Alabama, Jemison was raised in Chicago where she also went to high school. She attended Stanford University, earning a degree in Chemical Engineering in 1977, where she also took another degree's worth of courses in African and Afro-American Studies. In 1981, she graduated from Cornell with an M.D. and went on to intern at the Los Angeles County/USC Medical Center the next year.

In the early 1980s, Jemison worked as a Peace Corps Medical Officer, supervising medical care in West Africa and working on research related to vaccination with the Centers for Disease Control. When she returned to the United States in 1985, she began working again as a physician and applied to the astronaut program, taking extra engineering classes to boost her technical proficiency. She was selected in 1987 as part of Group 12. She flew aboard the Space Shuttle *Endeavor* in 1992 as a mission specialist on the Spacelab collaboration between the United States and Japan. Jemison spent just over 190 hours in space.[432]

A member of the backup crew for Jemison's STS-47 flight, Chiaki Mukai (born May 6, 1952) was the first Japanese woman to fly in space. Mukai was born in Tatebayashi, Gunma Prefecture in Japan. She studied at the Keio University School of Medicine to earn a medical degree in 1997, followed by a second doctorate in Physiology. In 1989, she became certified by the Japan Surgical Society as a cardiovascular surgeon and worked at a number of hospitals in Japan until 1985, when she was selected to be an astronaut by what was then called Japan's National Space Development Agency. Mukai flew on STS-65 and STS-95. Her career total of 566 hours of spaceflight included medical experiments conducted in orbit and support for research payloads including the Spartan spacecraft, which observed the sun.[433]

Women's participation in the scientific, technological, medical, and administrative aspects of spaceflight has been constant, if overshadowed, since its earliest days. In 1963, only days after Tereshkova's historic flight, the popular weekly photo magazine *Life* published a scathing editorial by Clare Boothe Luce on the apparent gap in achievement between the United States and the USSR in space. "Why did the Soviet Union launch a woman cosmonaut into space?" she asked. "Failure of American men to find the right answer to this question may

CHIAKI MUKAI

The first Japanese woman in space, physician Chiaki Mukai flew aboard the Space Shuttle *Endeavor* in 1992.

↓

MAE CAROL JEMISON

Physician and engineer Mae Carol Jemison was the first African American woman to fly in space, logging more than 190 hours aboard the Space Shuttle *Endeavor*.

yet prove to be their costliest Cold War blunder." In the wake of Tereshkova's flight, some people argued that it was just a publicity stunt—that "sex sells"— and the Russian space program was capitalizing on the spectacle of a woman in space. Luce dismissed these claims and instead argued the USSR's commitment to women cosmonauts reflected a larger commitment in socialist societies to the progress of women in general.

As a prominent conservative political commentator and wife of the magazine's publisher Henry Luce, Luce's argument was rooted in Cold War fears that communism was a threat because it might prove to better serve and make use of more of its people than capitalist societies. Luce framed the involvement of women in the space program, which had come to stand in for an existential conflict over the proper way to organize society, as an urgent barometer of the progress of the Cold War. According to Luce, Tereshkova's flight "symbolizes to Russian women that they actively share (not passively bask, like American women) in the glory of conquering space."[434]

If, as Luce suggested, women in spaceflight were an embodiment of the larger struggle of women in society, then their history offers an opportunity to reflect on the political stakes of women's involvement with science and technology in the twentieth century. Consistently marginalized in the high tech "man's world" of NASA's space programs, women were denied not only challenging careers in a nationally prestigious field but they also watched as their political position in American society was debated and dismissed with ease. With Cold War tensions abating in the 1970s and the subsequent collaborations between the United States and the USSR in spaceflight, women at last joined the astronaut corps in the United States. But the recovery of the stories of women whose contributions to spaceflight were irreplaceable but received far less public attention continues today.

Ellen Ochoa

(born May 10, 1958)

Ellen Ochoa is an American engineer and retired astronaut. Born in Los Angeles, California, Ochoa was the first Latina woman to fly in space as part of the crew of the shuttle *Discovery* in 1993. Ochoa attended San Diego State University as an undergraduate and earned a master's and doctorate in electrical engineering from Stanford. After receiving her PhD in 1985, Ochoa worked as a research engineer at Sandia National Laboratories in Albuquerque, New Mexico and NASA's Ames Research Center in Silicon Valley. Ochoa credits her mother, whose passion for learning kept her in part-time college courses all through Ochoa's childhood, as an important influence on her career.

In 1990, Ochoa was selected to astronaut candidacy as part of Group 13, a group of twenty-three NASA astronauts, and became an astronaut a year later. Her first spaceflight was aboard *Discovery* as a mission specialist and lasted nine days, in which the five-person crew conducted scientific experiments and deployed a research satellite to study the solar corona. A year later, Ochoa flew on *Atlantis* as the Payload Commander, spending ten days in space. On her 1999 flight on *Discovery*, the crew docked the shuttle for the first time with the International Space Station and delivered essential supplies and components ahead of the first human crews who would live there. Ochoa's last spaceflight was in 2002; the crew of *Atlantis* visited the ISS for eleven days and conducted spacewalks assisted by the station's robotic arm.

At the end of her flying career, Ochoa had logged nearly 1,000 hours in space. From 2012 to 2018, Ochoa served as the director of Johnson Space Center in Houston, Texas, only the second woman to head up NASA's human spaceflight headquarters. NASA has awarded Ochoa with its Distinguished Service Medal, Exceptional Service Medal, Outstanding Leadership Medal, and four Space Flight Medals.

CHAPTER 20

Reconfiguring the Female

DEVELOPING A PSYCHOLOGY OF WOMEN

"It is true that some scientists have categorically affirmed woman's inferior equipment," Leta Hollingworth conceded in her 1916 article, "Science and Feminism." But after years of putting scientists' claims about women's inferiority to the test, Hollingworth concluded that "they have voiced folk-lore and folk-ethics rather than science."[435] A psychologist trained at Columbia University, Hollingworth (May 25, 1886–November 7, 1939) dedicated much of her early research to dismantling psychology's theories about women's abilities and intellect. Effectively turning psychology's own tools against it, she and the feminist psychologists who followed her fought to expose how uncritically male psychologists had accepted cultural beliefs about women in order to prove and protect their own superiority in the field and in society at large.

Hollingworth belongs to what historians consider the first generation of women psychologists, who entered the field in the early days of its professionalization. In the 1890s, people in the field of psychology began to establish laboratories, professional societies, journals dedicated to the science, and university courses and departments. In the early decades of the twentieth century when the field was still comparatively new, more women worked in psychology than many other established sciences, such as chemistry and physics.[436] This did not mean, however, that psychology was wholly egalitarian. Male psychology PhDs, for instance, were more likely to find university and research positions than women with the same credentials. Instead they were often tacitly ushered into service work in schools, hospitals, and clinics or into

← Members of the New York Women's Liberation Army gather on a street corner to demand abortion rights during a demonstration in 1972.

lower-paid assistant positions.[437] Women were also much less likely to advance into leadership positions in universities and professional societies.

Women who entered the field during this time found themselves contending not only with psychology's professional discrimination but also the theories about women's inferior nature, theories that many male psychologists used to prop up that same discrimination. Psychologists adopted nearly wholesale nineteenth-century cultural beliefs about masculinity and femininity, which had been reinforced by scientists' claims that such beliefs were based in heredity and biology. Many theories posited that men alone were capable of intellectual thought and reason, whereas women were governed purely by emotion. This purportedly explained why men were capable of reaching intellectual greatness and social prestige while women were perfectly suited for domestic matters of child rearing and wifely duties.[438]

THE VARIABILITY HYPOTHESIS

In his 1871 *The Descent of Man*, Charles Darwin crystallized this view with the new science of evolutionary biology, claiming that man achieved "higher eminence, in whatever he takes up, than can woman—whether requiring deep thought, reason, or imagination or merely the use of the senses and hands." Stephanie Shields, a second-generation woman psychologist working in the twentieth century, has argued that the emphasis placed on "eminence" was crucial to this time period because social power was conceived as an inevitable outcome of one's abilities.[439] If women were not advancing in their careers in psychology, it would be argued that women simply were not made for that kind of high-level work and sophisticated research.

Outside of academia at the turn of the century, the first wave of feminism was fighting for suffrage amid increased attention to the status of women in society at large.[440] Many of the same theories of women's intellectual inferiority were used to justify women's disenfranchisement in politics and other aspects of society. So when women psychologists turned their attention to the gender biases baked into the field of psychology, they were also challenging the same biases that kept women from winning the right to vote and saw them relegated to the home as wives and mothers.

Part of Leta Hollingworth's drive to challenge science's views of women came directly from her experience as an unwilling housewife. Hollingworth, who was working as a school principal and teacher, was forced into a domestic role after she married her husband. Harry Levi Hollingworth was a psychology professor at

LETA HOLLINGWORTH

Leta Hollingworth was part of the first generation of women psychologists, and her work laid the foundation to overturn Victorian cultural beliefs about the psychology of women.

THE COMPARATIVE VARIABILITY OF THE SEXES AT BIRTH

HELEN MONTAGUE
New York City

AND

LETA STETTER HOLLINGWORTH
New York City

INTRODUCTION

The discussion of the comparative variability of the sexes began, somewhat vaguely, about a century ago, and bore on anatomical traits. The anatomist Meckel concluded, on pathological grounds, that the human female showed greater variability than the human male, and he thought that, since man is the superior animal, and variation a sign of inferiority, the conclusion was justified. Burdach and other anatomists declared the male to be more variable, and Darwin was led to conclude that among animals the male is more variable. Variation was now no longer regarded as a sign of inferiority, but as an advantage and a characteristic affording the greatest hope for progress. More recently greater mental variability has been inferred from alleged greater anatomical variability, and social significance has been attached by men of science to the comparative variability of the sexes. It has been stated that woman represents the static and conservative element in civilization, while man represents the dynamic and variable element—and that this accounts for the fact that nearly all historical achievement has been the achievement of men. It is further indicated that in the future, as in the past, in spite of any changes that may be wrought in the economic and social status of women, men will always lead women intellectually, because they are inherently more variable.

Prolonged reflection on this matter, and careful study of all available evidence lead to the conclusion that the data at present collected are inadequate for the formulation of any positive

335

The first page of Hollingworth's paper "The Comparative Variability of the Sexes at Birth," written with Helen Montague and published in 1914.

Columbia University and was supportive of her professional life, but schools in New York where she was seeking employment as a teacher would not hire married women.[441] After three years of unemployment, Hollingworth went back to school, enrolling as a graduate student in psychology at Columbia University in 1911. Columbia had opened its doors to women only eleven years before. After earning her master's degree, she continued into doctoral study under her mentor Edward Thorndike.

Thorndike and Hollingworth made for an unlikely mentor and mentee pair. Thorndike subscribed to psychological theories about women's inferiority that Hollingworth fully rejected, such as the Variability Hypothesis. This theory posited that men exhibited higher variation, or range, in physical and psychological traits while women were stuck in static mediocrity, never able to excel to a higher level. Thorndike described the implications of this theory, writing, "[I]f men differ in intelligence and energy by wider extremes than do women, eminence in and leadership of the world's affairs of whatever sort will

Psychologists adopted nearly wholesale nineteenth-century cultural beliefs about masculinity and femininity, which had been reinforced by scientists' claims that such beliefs were based in heredity and biology.

inevitably belong oftener to men. They will oftener deserve it."[442] Hollingworth used her own work to prove him wrong. With her colleague Helen Montague, Hollingworth examined 1,000 male and 1,000 female neonate hospital records, comparing physical traits such as birth weight and length. In their paper, "The Comparative Variability of the Sexes at Birth," Hollingworth and Montague found that female neonates exhibited a higher rate of physical variability than male neonates. In a followup paper of her own, Hollingworth laid out a point by point assault on the Variability Hypothesis, calling out Thorndike specifically and arguing that variation is a result of environment, not inheritance.[443] "No proof of greater male variability in mental traits can be found in the scant and inconclusive data available on the subject," she wrote. "The theory exists, but the evidence does not."[444]

For her dissertation under Thorndike, Hollingworth overturned the myth of mental deficiency in women brought on by menstruation. In "Functional Periodicity: An Experimental Study of the Mental and Motor Abilities of Women During Menstruation," Hollingworth detailed her study of male and female performance in a series of motor and cognitive ability tests. She found no evidence that women's performance decreased during menstruation. Despite their obvious differences, Thorndike could not deny the validity of Hollingworth's work. In fact, he was so impressed with her work that he offered her a job at the Columbia Teachers College and she accepted.

WOMEN PSYCHOLOGISTS IN TRANSITION

Hollingworth, along with many other first-generation women psychologists, laid the groundwork for the subfield "psychology of women." They took seriously the concerns of women and pushed back against psychology's uncritical acceptance of women's place in society as biological destiny. Psychology, however, did not change its views of women overnight, or even over the decades to come. Throughout the twentieth century, women psychologists would continue to encounter much of the same institutional and scientific discrimination faced by their first-generation predecessors.

While women in other professional fields and sciences often found more opportunities for work when wars created labor shortages, women psychologists found no such welcome. The Emergency Committee in Psychology was formed in 1940 under the purview of the National Research Council to lend their expertise to military matters. Women psychologists who hoped to contribute their expertise to the war effort were told to "be good girls and wait until plans could be shaped to include them."[445] Instead of waiting, a group of fifty women psychologists instead formed the National Council of Women Psychologists (NCWP). It marked the first time women psychologists organized with the purpose of professional advancement, and it was the only such group during wartime.[446] After the war, the committee dissolved and the organization was transformed into the International Council of Women Psychologists with a more general purpose of advancing practical applications

↑

NAOMI WEISSTEIN

Experimental psychologist Naomi Weisstein helped establish the
field of Feminist Psychology with her 1968 essay "Kinder, Küche,
Kirche as Scientific Law: Psychology Constructs the Female."

of psychology. While short-lived, the NCWP was essential in maintaining the urgency of the question of women's status in psychology between the first and second waves of feminism, the latter of which also heralded the entry of the second generation of women psychologists.[447]

The careers of this new cohort of women psychologists overlapped with the Civil Rights Movement, and, as a result, the second generation was more racially diverse than their white middle-class predecessors. The entry of more Black women psychologists into the field not only diversified the field but brought new perspectives to psychology's view of racial difference. Mamie Phipps Clark and her husband Kenneth (see page 213) were at the forefront of psychology's movement away from scientific racism to looking to social and environmental factors in racial differences.[448] Their work in this area helped prove that segregated education in the American South was harmful to Black children and provided critical evidence for striking down school segregation in the landmark Supreme Court case *Brown v Board of Education*. Jeanne Spurlock (July 19, 1921–November 25, 1999), who served as Deputy Medical Director of the American Psychiatric Association for seventeen years, focused on psychological development in Black children. Like Clark, she used her expertise to advocate for Black children, lobbying legislators to provide mental health funding for minority communities.

With renewed interest generated in part by second-wave feminism, this second generation again took up the psychology of women, led by experimental psychologist Naomi Weisstein (October 16, 1939–March 26, 2015). A talented researcher, Weisstein graduated top of her 1964 class at Harvard in just two and a half years, despite one of her professors telling her that "[w]omen don't belong in graduate school."[449] She also held a National Science Foundation postdoctoral fellowship in mathematical biology at the University of Chicago. Yet, early in her job search, she realized how common her Havard professor's views were and how difficult it was for a woman, even one as decorated as her, to find a job in the field. In her essay, "How can a little girl like you teach a great big class of men?", Weisstein recalled continuous dismissive encounters with male colleagues on the job market. One asked, "Who did your research for you?", and another admonished her, "You ought to get married."[450]

CREATING A FEMINIST PSYCHOLOGY

Weisstein understood that the discriminatory behaviors of male psychologists were not limited to academia and could have damaging implications for the larger cause of women's liberation. "[W]hen we are about to consider the liberation of women, we naturally look to psychology to tell us what 'true' liberation would mean: what would give women the freedom to fulfill their own intrinsic natures," Weisstein wrote in her 1968 essay, "Kinder, Küche, Kirche as Scientific Law: Psychology Constructs the Female." But Weisstein cautioned that what psychologists described as the nature of women more often reflected

Women psychologists asked different questions and sought answers in new ways and places that their male colleagues could not or would not.

male fantasies about women, rather than what women themselves want to be. In the essay, she called out prominent contemporary psychologists, like Bruno Bettelheim of the University of Chicago, who had said that as much as women might want to be scientists "'they want first and foremost to be womanly companions of men and to be mothers.'"[451]

The willingness with which psychology embraced "sexist norms of our culture," Weisstein said, had rendered the field all but worthless when it came to the liberation of either women or men. "Psychology has nothing to say about what women are really like, what they need and what they want, especially because psychology does not know," she contended. Much like Hollingworth decades before, Weisstein demonstrated how psychology had constructed theories about women without evidence, mistakenly relying too heavily on innate sex differences rather than social and cultural context. Without evidence and a careful consideration of social context, Weisstein argued that psychology was incapable of knowing the lived experiences of women.

Weisstein's paper resonated with an anger and acerbic wit not typically found in scholarly writing. It is not difficult to understand why she opted for a salvo rather than a clinical approach to reiterate arguments that had been made by Hollingworth's generation nearly a half century before. The impact of her paper was immediate; it was reproduced in more than thirty different disciplinary readers, ranging from philosophy of science to political science, and was canonized in second-wave feminist literature in the 1970 publication *Sisterhood is Powerful: An Anthology of Writings from the Women's Liberation Movement*.[452] Weisstein ushered in a new era of research in the psychology of women that we now call feminist psychology.

In 1973, the American Psychological Association (APA), the largest professional body of psychologists in the United States, incorporated feminist psychology into its organization. The Council of Representatives approved a petition to install the Division of the Psychology of Women (Division 35), which Weissten helped form. Now called Society for the Psychology of Women, the Division's express purpose was "to promote the research and the study of women [and] to encourage the integration of this information about women with the current psychological knowledge and beliefs in order to apply the gained knowledge to the society and its institutions."[453] By obtaining institutional status within the APA, women psychologists and the field of feminist psychology were granted the benefits of a sponsoring journal, slotted program time at the APA's annual meetings, and representation on the Council of Representatives.[454] Their Division and journal, *Psychology of Women Quarterly*, are still running today. Since its founding, the Division has expanded to include division sections that better reflect the diversity of women's experiences: Section I: Psychology of Black Women; Section III: Concerns of Hispanic Women/Latinas; Section IV: Lesbian, Bisexual, and Transgender Concerns; Section V: Psychology of Asian Pacific American Women; and Section VI: Alaska Native/American Indian/Indigenous Women.

A generation prior to Division 35, Hollingworth wryly quipped, "It is amusing to note how every sex difference that has been discovered or alleged has been interpreted to show the superiority of males."[455] With the formulation of feminist psychology still decades away, Hollingworth was very much aware of the necessity for a more inclusive psychology, both in its practitioners and in its theory. Psychology's views of male superiority and female inferiority had remained unchallenged and stuck in the nineteenth century, but the inclusion of more women psychologists heralded new interpretations of the differences between men and women. What is more, feminist psychologists have brought issues of domestic and sexual violence, societal pressures, and systemic oppressions like racism and homophobia under the purview of the discipline. Women psychologists asked different questions and sought answers in new ways and places that their male colleagues could not or would not.

CHAPTER 21

The Problem with "Female Firsts"

BEYOND MARIE CURIE

When we consider the history of science from the earliest days of antiquity to our technoscientific present, where and how women contributed to this most important human intellectual project is often not apparent in popular retellings and representations. The names of men who have reshaped our knowledge of nature are much more familiar, commonplace figures in our collective sense of the march of scientific progress. But the stories of the women who are as integral to this story are understudied because many of them were never permitted to participate in the mainstream of science that historians most often take as their subject. Women were barred from colleges and universities, from the lecture and surgical theaters of the great halls of learning, and from the discourse that we look to for understanding about the history of science. Theirs was a constant clawing at the edges of spaces where they were not permitted, where their very interest in the investigation of nature was often seen as distasteful or suspicious, and where knowledge of the world seen as unfit for feminine sensibilities. When we look to the rolls of these institutions for the contributions, we will always be disappointed.

But another force underlies the marginalization of women in science. Our contemporary imagination of women scientists is dominated by a few luminous figures, whose stories are trotted out at every conceivable opportunity in due deference to women in science. Marie Curie (November 7, 1867–July 4, 1934), for instance, does not appear elsewhere in this book, not because she was not important but since she is the single most well-studied woman scientist in history and the deserved light in which we now see her has obscured nearly

← A magnetar, a type of extreme pulsar, captured by NASA's Chandra X-ray Observatory.

all other women scientists. This phenomenon of "firsts" is a consequence of otherwise well-meaning attempts to recover the history of women in science, by looking in the same places we would look for men. But, as we have shown in this book, the female "firsts" in the history of science represent only a tiny minority of those who were able to force their way into the scientific mainstream. One final example may serve to underscore this point and to demonstrate the ways that certain scientific disciplines—those that are ancient, codified, and whose subjects are particularly captivating to we ordinary observers—shed the same kind of obscuring light as deified figures like Curie have come to do.

DARK MATTER

Vera Rubin (July 23, 1928–December 25, 2016), one of the most influential astronomers of the twentieth century, was born Vera Florence Cooper in Philadelphia in 1928. She was educated at Vassar, where the lauded astronomer Maria Mitchell (August 1, 1818–June 28, 1889) was a professor in the nineteenth century, and Cornell—after she was turned away from Princeton for being a woman.[456] During her years at Vassar just after World War II, Rubin met her husband Robert, worked summers at the Naval Research Laboratory in Washington D.C., and served as "clock winder, paper grader, and telescope helper" for a professor.[457] When she and Robert moved to Ithaca, New York, in 1948, she began graduate work at Cornell where she learned from famous physicists including Richard Feynman and Hans Bethe and worked on her master's thesis on the velocity distribution of galaxies under astronomer Martha Stahr Carpenter.[458] Rubin later recalled a vivid memory of her department chair offering to present her master's research at the 1950 meeting of the American Astronomical Society (AAS), suggesting that since she would have a small child with her and obviously be unable to attend, he could present the research provided he attached his own name. Rubin refused and presented her own work to the AAS, with significant pushback from the AAS attendees. *The Washington Post* reported on the conference the next day with the headline "Young Mother Figures Center of Creation by Star Motions." The article reported that the response to her paper was "polite" if "persistent," but Rubin remembers that day much differently.[459] "One by one," she wrote in an autobiographical sketch from 2012, "many angry sounding men got up to tell me why I could not do 'that.'"[460]

Rubin went to Georgetown in 1952 to pursue her doctoral studies, extending her master's research on galaxies, and she published her dissertation on the *Proceedings of the National Academy of Sciences*. She stayed on at Georgetown

↑

VERA RUBIN

Astronomer Vera Rubin's work on the rotation of galaxies
provided some of the first evidence for dark matter,
a major component of the structure of the universe.

for ten years teaching and researching, and in the early 1960s, she began a series of observations at various telescopes around the country, including Kitt Peak National Observatory in Arizona. In 1965, she made an observing trip to Caltech's Palomar Observatory before women were officially permitted to use the facilities.[461] Her friend Margaret Burbidge had done the same, using her husband Geoff's privileges to get a little time on the telescope.[462]

In 1965, Rubin took a research job at the Carnegie Institution's Department of Terrestrial Magnetism to focus on observation, where she was forced to accept a decreased salary in order to be home in the afternoons with her children. She began collaborating with Kent Ford on the spectra of galaxies, working first on imaging the Andromeda Galaxy at the Lowell Observatory.[463] Rubin determined that unlike our own galaxy, in which stars positioned further from the center rotate more slowly, the rotation curve in Andromeda's galaxies was flat. In order for these observations to be consistent with Newtonian physics, something had to be contributing enough mass to flatten the curve. This was the first evidence for what we now call dark matter, a mysterious, invisible category of matter that makes up most of the known universe.[464]

Even where women have made inarguable contributions to our knowledge of nature, of the kind that even the most disinterested observer cannot overlook, they have still faced systemic discrimination at the highest levels of science. Like Rubin, astrophysicist Jocelyn Bell Burnell (born July 15, 1943) made an important cosmological discovery, rewriting the very map of the heavens, and she did so from a place of professional marginalization.

THE NAMING OF A NEW STAR

Burnell is an astrophysicist, best known for discovering pulsars, a type of rapidly rotating star. Born in 1943 in Northern Ireland, Burnell went to college at the University of Glasgow where she earned a physics degree, and later in 1969, she received her doctorate from the University of Cambridge.[465] At Cambridge, while working on her dissertation, Burnell made her famous pulsar discovery. Working with her advisor Antony Hewish, their research group built a new kind of radio telescope to better study quasars, the cores of galaxies. Radio telescopes, unlike the optical telescopes used by Rubin, collect radio signals from distant celestial objects. The one Hewish and Burnell built was 4.5 acres across, and construction began in 1967.[466]

Burnell was in charge of operating the telescope and monitoring the ninety-six feet (nearly thiry meters) of chart paper the instrument generated each day, which visually documented the radio waves received by the telescope. Analysis of these charts was done visually and by hand, something else Burnell was responsible for, and it was in the flowing wave forms of the chart paper that she first discovered what she called "scruff."[467] Part of the job entailed visually identifying signals from distant objects like stars and more local interference from earth or the solar system, so the "scruff" stood out to Burnell

---

Our contemporary imagination of women scientists is dominated by a few luminous figures, whose stories are trotted out at every conceivable opportunity in due deference to women in science.

---

because it was a very regular pattern. Hewish reviewed the signal and initially concluded that because of its regularity, it was likely made by humans and was just interference. But they still needed to isolate the pattern to determine its source and prevent it from contaminating other data. Burnell was not entirely convinced, however, that the pattern was down to human interference, and later she jokingly said of the discovery that "due to a truly remarkable depth of ignorance I did not see why they could not be from a star."[468]

Burnell was right; the signal *was* from a star, a kind that had never been described before. In 1968, she and Hewish published a paper on their findings, but the source had yet to be clearly identified. Because they could not disprove that the source was some kind of extraterrestrial intelligence, though they thought that highly unlikely, the possibility was raised and seized upon by the press. Burnell remembered being disappointed that her dissertation project was potentially being hijacked by the much bigger discovery of "little green men."[469] The paper, "Observation of a Rapidly Pulsating Radio Source," proposed that the signals came from "oscillations of white dwarf or neutron stars," which was later confirmed.[470] In 1974, Hewish and Martin Ryle shared

↑

JOCELYN BELL BURNELL

As a postgraduate student in 1967, Jocelyn Bell Burnell discovered
the first radio pulsars, a major breakthrough in astronomy for which
her advisor was awarded a Nobel Prize.

the Nobel Prize in Physics. Hewish was credited with "his decisive role in the discovery of pulsars."[471]

THE PHENOM FIRSTS

Both Rubin and Burnell are frequently featured on lists of unsung women in science, in part because both are often represented as being unfairly passed over for Nobel Prizes, the most prestigious award for scientists. Burnell has long maintained that not being included with Hewish in the 1974 prize has never bothered her. But in a 2004 article "So Few Pulsars, So Few Females," she acknowledged that the Nobel committee's decision was not entirely fair, writing,

> *"Arguably, my student status and perhaps my gender were also my downfall with respect to the Nobel Prize, which was awarded to Professor Antony Hewish and Professor Martin Ryle. At that time, science was still perceived as being carried out by distinguished men leading teams of unrecognized minions who did their bidding and did not themselves contribute other than as instructed!"*[472]

Astronomy in particular is an attractive field for debate about the exclusion of women from science and the misattribution of their accomplishments in awards and professional recognition. Both Rubin and Burnell were responsible for the type of discovery we often find most impressive and most miraculous—the naming of new taxa of our universe. The distant and mysterious phenomena their research illuminated captivates our imaginations and traces the remote edges of human knowledge. But even within their own field, let alone the larger history of science, they are hardly "firsts" in the conventional sense. Carolyn Herschel at the turn of the nineteenth century discovered a number of comets, and Rubin and Burnell's contemporary Carolyn Shoemaker repeated the feat some 200 years later. Burnell recognized the company that she kept, writing that "[t]here have been some excellent female astronomers in the past who were not fully recognized for their contributions."[473]

FORCES OF NATURE

The gold medals given to Nobel Laureates in Physics and Chemistry and Physiology or Medicine are curious artifacts of the history of science. Unlike the medals for Peace or Economics, which bear the likeness of Alfred Nobel, or Literature which shows a young man receiving the gifts of the muse, the science medals feature only female figures. Rendered in relief on the Physics and Chemistry medal is an allegorical scene of Nature being revealed by the Genius of Science. The latter is a female figure grasping in one hand a scroll and the other the veil she lifts away from the face of Nature, who is bare-breasted and holding a cornucopia.[474] These allegorical figures appear throughout the visual history of Western science, from the frontispieces of scientific books to sculptures to the pediments of columned Greek style public buildings. As historian Londa Schiebinger has observed, the image of woman has had pride of place in the most celebrated scientific spaces, but often only as the mythological embodiment of it highest ideals. The real women of science are much more difficult to find, in part because they are often denied access to the very institutions that bear their image in such esteem.

We know that women have *always* pursued knowledge of nature, so the task that falls to us now is explaining and understanding the structural and institutional forces that have hindered women's quest and submerged them beneath the stream of history. Women have faced not only institutional barriers—prohibition against education for women, bans on women being named to important scientific societies—but also everyday sexism, harassment, abuse, and in some cases violence. A richer, fuller understanding of the history of science is gained by dismissing the question of *if* women have been capable of scientific achievement and instead seeking out and explaining the forces that have hidden them from view in the hope of creating better, more inclusive institutions for future women in science.

Afterword

Other women to inspire

Elizabeth Garrett-Anderson
(1836–1917, top right)

Heavily influenced by American physician Elizabeth Blackwell, Elizabeth Garrett-Anderson took up a life in medicine instead of marriage. Anderson struggled for years to find a medical school that would accept her, ultimately earning her medical degree in Paris. She became the first British woman to qualify as a physician and surgeon. She later founded the New Hospital for Women in London and was appointed dean of the London School of Medicine for Women, which she also helped establish. In 1908, Anderson added another "first" to her name when she became mayor of Aldeburgh and first woman mayor in England.

Sophia Jex-Blake
(1840–1912)

Sophia Jex-Blake was influential in opening the medical field to women physicians. Denied study of medicine by both American and English universities, Jex-Blake and six other women, known as the Edinburgh Seven, enrolled at the University of Edinburgh, but she was notallowed to earn a degree. She spearheaded the Medical Act of 1878, which struck down previous Acts that excluded women from medical licensing. Finally, in 1877, Jex-Blake graduated as a M.D. from the University of Berne.

Aletta Henriëtte Jacobs
(1854–1929, center right)

Aletta Jacobs—physician, suffragist, birth control advocate, peace activist— was the first woman in The Netherlands to attend a university and receive a medical degree. Despite resistance from the medical community, Jacobs opened the world's first birth control clinic. In serving women at her clinic, she combined her advocacy for women's healthcare and for women's rights in labor, seeing firsthand the toll long work days in deleterious conditions took on women's bodies. Beyond medicine, Jacobs co-founded the Woman Suffrage Alliance and helped establish the Women's International League for Peace and Freedom.

Margaret Alice Murray
(1863–1963, below left)

Archaeologist, Egyptologist, folklorist, and anthropologist, Margaret Murray left her mark on multiple fields over the course of her 100-year-long life. Murray was the first woman lecturer in archaeology in the United Kingdom and established the first two-year universiting training program for future field workers at University College, London. As a folklorist, she developed the "witch cult hypothesis," the earliest anthropological study on witchcraft in Britain, which has since become the foundation of scholarly studies of witchcraft. A towering figure as a "first" in her field, she served as a mentor to all those who passed through her classroom.

Gertrude Bell
(1868–1926)

Dubbed "Queen of the Desert" and the "Female Lawrence of Arabia," Gertrude Bell was an archaeologist, spy, and diplomat. Bell traveled throughout the Middle East, excavating ruins in Syria and mapping large portions of the region. During World War I, Bell assisted British Intelligence by escorting soldiers across the deserts of the Middle East, and she became a knowledgeable intelligence officer, trusted by both the British and Arabian people. Her most enduring legacy is her diplomatic role in placing the Hāshimite dynasty on the throne of Iraq, a move that helped build the modern-day country. First an archaeologist, Bell cherished her work in preserving Iraq's antiquities the most. Insisting that Iraq's antiquities remain in Iraq, Bell established the Baghdad Antiquities Museum, now the National Museum of Iraq, the seat of Iraq's cultural heritage.

Lise Meitner
(1878–1968, below right)

Physicist Lise Meitner, the first woman professor of physics at University of Berlin, was part of the team that discovered nuclear fission, the discovery that led to the creation of the atomic bomb. While seeking refuge from the Nazis in Sweden, Meitner and her nephew Otto Frish described the physical process of fission and gave the process its name. Despite her integral role in the discovery, she was excluded from the Nobel Prize, which went only to Otto Hahn. Meitner refused to take part in the application of her research to create nuclear weapons, and for this reason, she is remembered to this day by her gravestone epitaph written by Frisch: "a physicist who never lost her humanity."

Janaki Ammal
(1897–1984)

Janaki Ammal's skill as a plant cytologist was matched only by her passion for the native flora of India. The first Indian woman to receive a doctorate in botany in the United States, Ammal excelled in breeding interspecific and intergeneric hybrids, and her work on sugarcane hybrids resulted in the first sweet variety native to India. As director of the Central Botanical Laboratory, Ammal advocated for the preservation of India's flora and for the prioritization of Indians' knowledge about the Indigenous plants over

that of British botanists. In the 1970s and 80s, Ammal joined the Silent Valley Movement, one of the most significant environmental projects of the time that successfully advocated for the preservation of the Silent Valley's evergreen tropical forest, which is now a national park.

Wangarĩ Muta Maathai
(1940–2011, centre left)

Wangarĩ Muta Maathai was trained in biological sciences but made a name for herself as an environmental crusader and political activist. In 1977, she founded the Greenbelt Movement in Kenya, an NGO that tied together women's rights, reforestation, and environmental conservation. She was an elected member of Kenya's National Assembly and Assistant Minister of environment, natural resources, and wildlife. In 2004, she was awarded the Nobel Peace Prize for a career in environmental development that protected both democracy and human rights; she was the first African woman to receive the prize.

Octavia Hill
(1838–1912)

Octavia Hill, activist and social reformer, spearheaded the British open-space movement and improved housing and living conditions for Britain's working poor. Hill acquired and renovated properties with extended space and sanitation services, and portions of the rent went back to the community for shared playgrounds and classrooms. With deep conviction that England's land belonged to the people, she advocated for opening up public space and protecting what open space remained in the countryside. In 1895, Hill helped found the National Trust, which today comprises millions of members who oversee 600,000 acres of countryside in the United Kingdom and hundreds of miles of coastline.

Kalpana Chawla
(1961–2003, below center)

Born and raised in Karnal, India, Kalpana Chawla was the first Indian woman to fly in space and the first woman to study aeronautical engineering at Punjab Engineering College. Chawla was selected by NASA as an astronaut in 1994, and in 2003, she served as the mission specialist and robotic arm operator aboard the space shuttle *Columbia* for the STS-87 mission. After a fifteen-day mission in space, *Columbia* broke apart upon re-entry, killing Chawla and her six-crew members. Chawla was posthumously awarded the Congressional Space Medal of Honor, the NASA Space Flight Medal, and the NASA Distinguished Service Medal.

Samantha Cristoforetti
(1977–present, top center)

Samantha Cristoforetti is currently an astronaut with the European Space Agency and a captain in the Italian air force. In 2009, Cristoforetti along with five others were selected from an application pool of 8,000 to join the ESA astronaut corps. Serving on the mission Futura, Cristoforetti spent 199 days in space, the longest spaceflight of any European and the longest single mission for any woman. After several years leading a student-centered team focused on future Moon missions, Cristoforetti currently serves as crew representative for the ESA's Lunar Orbital Platform – Gateway project.

Maryam Mirzakhani
(1977–2017)

In 2014, Maryam Mirzakhani became the first woman and the first Iranian to receive the Fields Medal, the most prestigious award in mathematics granted every four years to mathematicians under the age of forty. A professor at Harvard University, Mirzakhani specialized in hyperbolic geometry, moduli spaces, Ergodic theory, Teichmüller theory, and symplectic geometry. Her work was highly theoretical, but has practical applications in understanding the origins of the universe and in engineering and material sciences.

Marie Skłodowska Curie
(1867–1934, top left)

As the first woman to win a Nobel Prize and the only woman still to have won the award in two different fields (physics and chemistry), Marie Skłodowska Curie is known around the world for her co-discovery of polonium, radium, and the phenomenon of radioactivity. Less known but no less important was Curie's invention of the "radiological car," a vehicle equipped with an X-ray machine and darkroom equipment, which she developed for the World War I battlefield. She and her daughter Irene taught 150 women to operate the equipment, who all together delivered X-ray exams to more than 1 million soldiers. Much more than the enigmatic woman in black working in a lab, Curie was a complex person who embraced the practical application of her skills and her science.

Rosalind Franklin
(1920–1958, centre middle)

Sometimes called the "Dark Lady of DNA," Rosalind Franklin is best known for her "Photo 51," the X-ray image of DNA that led James Watson and Francis Crick to determine DNA's double-helix structure. Her role as the woman slighted by Watson and Crick in the search for the double helix has completely eclipsed her pioneering career in crystallography, for which she was renowned in her own time. She used X-ray crystallography to study both plant and animal viruses, including polio, and in the process, she developed our understanding of how virus RNA behaves. Still today, many virologists regard her X-ray photographs as some of the most beautiful ever taken.

Index

Afterword

Acknowledgments

We would like to thank several people for their help with this book. Nathan Kapoor, Cornelia Lambert, Robert Davis, and Asif Siddiqi for their assistance in tracking down hard-to-find sources. Ashley Roland and Brooke Steiger for guidance and technical help. We are extremely grateful to Kathleen Sheppard and Deanna Day for reading the entire manuscript and providing insightful and essential comments and feedback. We are additionally indebted to Lydia Pyne for walking us through the intricacies of publishing and providing encouragement every step of the way. Our team at Lady Science kept the lights on while we worked on this project and inspire us with their commitment to sharing this history.To our families, we thank you for all the support you've given us throughout this project, especially Erik for going to get beer while we were in the throes of editing the manuscript. And, of course, we extend our deepest gratitude to the historians who did all the heavy lifting to recover from history the lives of many of the women included in this book and on whose work this book profoundly depended. A big thank you to everyone at Quarto Group and White Lion Publishing for asking us to work on this project and for shepherding this book from start to finish.

Picture Credits

Endnotes

INTRODUCTION

1. Betty De Shong Meador, *Inanna, lady of largest heart: poems of the Sumerian high priestess Enheduanna* (Austin: University of Texas Press, 2000), 52–53.

2. This object is in the collection of the Penn Museum and images of both the original and the restored disk have been digitized: *https://www.penn.museum/ collections/object/293415*

3. Charles Keith Maisels, *The Emergence of Civilization* (London, Routledge, 1990), 121. Quoted in *Meador, Inanna, lady of largest heart*, 37.

4. Meador, *Inanna, lady of largest heart*, 37.

5. Marilyn Ogilvie and Joy Harvey, eds., *The Biographical Dictionary of Women in Science*, Volume 1 (New York: Routledge, 2000), 638.

6. Ibid.

7. Margaret Gaida, "Muslim Women and Science: The Search for the "Missing" Actors," *Early Modern Women* 11, no. 1 (Fall 2016): 199.

8. Ibid., 202.

9. Ibid., 205.

CHAPTER 1

10. J. F. Nunn, Ancient Egyptian Medicine (Norman: University of Oklahoma Press, 1996), 124.

11. Gay Robbins, *Women in Ancient Egypt* (London: British Museum Press, 1993), 116.

12. Robbins, *Women in Ancient Egypt*, 79.

13. Ibid., 80.

14. Ibid., 82.

15. Nancy G. Siraisi, *Medieval and Early Renaissance Medicine: An Introduction to Knowledge and Practice* (Chicago: The University of Chicago Press, 1990), 1.

16. Ibid., 2.

17. Ann Ellis Hanson, "Diseases of Women 1," *Signs* 1:2 (1975): 570.

18. Ibid., 572.

19. Ibid., 573.

20. Valerie French, "Midwives and Maternity Care in the Roman World," in *Midwifery and the Medicalization of Childbirth: Comparative Perspectives*, ed. Edwin van Teijlingen, George Lowis, Peter McCaffery, and Maureen Porter (New York: Nova Science Publishers, 2004), 54.

21. Ibid., 55.

22. Ibid., 56.

23. Ibid.

24. Ibid., 54.

25. Charlotte Furth, *A Flourishing Yin: Gender in China's Medical History: 960–1665* (Berkely: University of California Press, 1999), 48.

26. Ibid., 268.

27. Ibid., 276.

28. Ibid., 270.

29. Ibid., 277.

CHAPTER 2

30. Oliver Phillips, "The Witches' Thessaly" in *Magic and Ritual in the Ancient World* 141, eds. Paul Mirecki and Marvin Meyer (Brill: 2002), 379-380.

31. D.E. Hill, "The Thessalian Trick," *Rheinisches Museum für Philologie* 116 (1973): 223. Translations by Hill.

32. Plato, *Gorgias*, trans. Benjamin Jowett (ebooks@Adelaide, 2014) *https://ebooks.adelaide.edu. au/p/plato/p71g/complete.html*.

33. Hill, "Thessalian Trick," 225.

34. Ibid., 225.

35. Plutarch, "Conjugalia Praecepta" in *Moralia*, trans. Frank Cole Babbitt (Harvard University Press, 1928) *http://www.perseus.tufts.edu/hopper/text?doc= Perseus%3Atext%3A2008.01.0181%3Asection%3D48*.

36. Richard B. Stothers, "Dark lunar eclipses in classical antiquity," *Journal of the British Astronomical Association* (1986): 95; Ovid, *Metamorphoses* (B. Law, 1979), 63.

37. Bernard R. Goldstein, "*Babylonian Eclipse Observations from 750 BC to 1 BC*, edited by Peter J. Huber and Salvo De Meis," review of Babylonian Eclipse Observations from 750 BC to 1 BC, by Peter J. Huber and Salvo De Meis, Aestimatio 1, 2004, 123.; J.M. Steele and F.R. Stephenson, "The Accuracy of Eclipse Times Measure by the Babylonians," *Journal for the History of Astronomy* (1997).

38. Peter Bicknell, "The Witch Aglaonice and Dark Lunar Eclipses in the Second and First Centuries BC," *Journal of the British Astronomical Association*, no. 93 (1983): 160.

39. Quoted in Fiona Maddocks, Hildegard of Bingen: The Woman of her Age (London: Doubleday, 2001), 54-55.

40. Ibid., 56.

41. Ibid., 59.

42. Ibid., 60.

43. Marsha Newman, "Christian Cosmology in Hildegard of Bingen's Illuminations," Logos: A Journal of Catholic Thought and Culture 5, no. 1 (Winter 2002): 42.

44. Sharon Jones and Diana Neal, "Negotiable Currencies: Hildegard of Bingen, Mysticism and the Vagaries of the Theoretical, Feminist Theology 11, no. 3 (2003): 379.

45. Barbara Newman, Sister of Wisdom: St. Hildegard's Theology of the Feminine (University of California Press, 1987), 29.

46. Amy Hollywood, "'Who Does She Think She Is?': Christian Women's Mysticism," Theology Today, no. 60 (2003): 8.

47. Newman, Sister of Wisdom, 29.

CHAPTER 3

48. Swerdlow, N. M. "Urania Propitia, Tabulae Rudophinae Faciles Redditae a Maria Cunitia Beneficent Urania, the Adaption of the Rudolphine Tables by Maria Cunitz." A Master of Science History: Essays in Honor of Charles Coulston Gillispie, vol. 30, Springer, 2012, pp. 81–121.

49. Beneficent Urania, the Adaption of the Rudolphine Tables by Maria Cunitz," 81.

50. First law: planets move in elliptical orbits with the sun as the foci. Second law: the center of the sun and the center of a planet sweep out equal area in equal intervals of time. Third law: the amount of time a planet takes to orbit is directly related to its distance from the sun.

51. Swerdlow, 81.

52. Londa Schiebinger, The Mind Has No Sex?: Women in the Origins of Modern Medicine (Harvard University Press, 1989), 37.

53. Ibid., 44.

54. Grier, When Computers Were Human, 13–14.

55. Ibid., 14–15.

56. All information on Newton and Halley comes from Grier, When Computers Were Human, 13–15.

57. Ibid., 20.

58. Whaley, "Women's History as Scientists: A Guide to the Debates," 135.

59. Bernardi, "The comet calculator: Nicole-Reine Lepaute, Cosmos Magazine," 2018.

60. Roberts, "Learned and Loving: Representing Women Astronomers in Enlightenment France," 21.

61. Qutd. in Roberts, 24.

62. Roberts, 15.

63. Swerdlow, 85.

64. Mazzotti, Massimo. The World of Maria Gaetana Agnesi, Mathematician of God (Johns Hopkins University Press, 2007), 15.

65. Massimo Mazzotti, "Maria Gaetana Agnesi: Mathematics and the Making of the Catholic Enlightenment." Isis, vol. 92, no. 4, 2001, 670.

66. Ibid., 673.

67. Ibid., 674.

68. Agnesi, Maria Gaetana. Analytical Institutions in Four Books: Originally Written in Italian. Translated by Rev. John Colson, Taylor and Wilks, (London, 1801), XVIII

69. Ki Che Leung, Biographical Dictionary of Chinese Women, 231.

70. Benjamin A. Elman, A Cultural History of Modern Science in China, 17.

71. On the Tychonic system in Qing China see Elman pg 4. On Zhenyi's helio-centric thesis see Peterson,

Notable Women of China: Shang Dynasty to the Early Twentieth Century, 344–345.

72. Peterson, 344.

CHAPTER 4

73. As a loving assistant see Londa Schiebinger, The Mind Has No Sex?: Women in the Origins of Modern Medicine (Harvard University Press, 1989), 261; as collaborator see Meghan K. Roberts, "Philosophes Mariés and Espouses Philsophiques: Men of Letters and Marriage in Eighteenth-Century France," French Historical Studies 35, no. 3, (Summer 2012): 536.

74. Patricia Fara, Pandora's Breeches: Women, Science and Power in the Enlightenment (Pimlico, 2004), 176.

75. Fara, Pandora's Breeches, 176.

76. Ibid., 178.

77. Ibid., 178.

78. Ibid., 173.

79. William B. Ashworth, "Scientist of the Day – Elisabeth Hevelius," published by the Linda Hall Library December 22, 2017. https://www.lindahall.org/elisabeth-hevelius/.

80. Fara, Pandora's Breeches, 137

81. Joseph L. Spradley, "Two Centennials of Star Catalogs Compiled By Women," The Astronomy Quarterly 7 (1990): 178.

82. Gabriella Bernardi, "Elisabetha Catherina Koopman Hevelius (1647–1693)" in The Unforgotten Sisters (Springer International Publishing, 2016), 70.

83. Spradley, "Two Centennials," 178.

84. Michelle DiMeo, "'Such a sister became such a brother': Lady Ranelagh's influence on Robert Boyle," Intellectual History Review 25, no.1 (2015): 23.

85. Ibid., 23.

86. Ibid., 29.

87. Ibid., 29.

88. Alan Cook, "Ladies of the Scientific Revolution," Notes and Records: The Royal Society Journal of the History of Science 51, no. 1 (1997): 2.

89. Caroline Herschel, Memoir and correspondence of Caroline Herschel, ed. Mrs. John Herschel (London: John Murray, 1879), ix.

90. Richard Holmes, The Age of Wonder (Vintage Books: 2008), 63–67; Michael Hoskin, "Caroline Herschel's Life of 'Mortifications and Disappointments,'" Journal for the History of Astronomy 45, no. 4 (2014): 443–445.

91. Herschel, Memoir and Correspondence, 52.

92. Herschel, Memoir and Correspondence, 76.

93. Keiko Kawashima, "The Evolution of the Gender Question in the Study of Madame Lavoisier," Historia Scientiarum 23, no. 1 (2013): 33.

CHAPTER 5

94. Katharine Park, *Secrets of Women: Gender, Generation and the Origins of Human Dissection* (Cambridge, Zone Books, 2010).

95. Rebecca Messbarger, "Waxing Poetic: Anna Morandi Manzolini's Anatomical Sculptures," *Configurations 9*, no. 1 (2001): 66-68.

96. Ibid., 70-74.

97. Rebecca Messbarger, *The Lady Anatomist: The Life and Work of Anna Morandi Manzolini* (Chicago: University of Chicago Press, 2010), 10-11.

98. Ibid., 11-12. See also Marilyn Ogilvie and Joy Harvey, eds., *The Biographical Dictionary of Women in Science: Pioneering Lives from Ancient Times to the Mid-20th Century*, Volume 1 (New York: Routledge, 2000), 426.

99. Ibid., 13.

100. Messbarger, "Waxing Poetic," 74.

101. Ibid., 76.

102. Ibid., 85.

103. Quoted in Rose Marie San Juan, "The Horror of Touch: Anna Morandi's Wax Models of Hands," Oxford Art Journal 34, no. 3 (2011): 439.

104. Ibid., 66.

105. Messbarger, The Lady Anatomist, 11.

106. See also Chapter 4, this volume.

107. Londa Schiebinger, *The Mind Has No Sex? Women in the Origins of Modern Science* (Cambridge: Harvard University Press), 1989, 246.

108. Ibid.

109. Ibid., 250.

110. Ibid., 251.

111. Ogilvie and Harvey, *Biographical Dictionary*, 426.

112. Schiebinger, *The Mind Has No Sex?*, 251.

113. Ibid., 252.

114. Ibid.

115. Ibid., 255.

CHAPTER 6

116. Londa Schiebinger, *Plants and Empire: Colonial Bioprospecting in the Atlantic World* (Cambridge, MA, Harvard University Press), 50.

117. Glynis Ridley, *The Discovery of Jean Baret: A Story of Science, the High Seas, and the First Woman to Circumnavigate the Globe* (Broadway Books, 2011).

118. Schiebinger, *Plants and Empire*, 46.

119. Schiebinger, *Plants and Empire*, 51.

120. Sandra Knapp, "The plantswoman who dressed as a boy," Nature 470 (2011) *https://www.nature.com/articles/470036a*.

121. Schiebinger, *Plants and Empire*, 50.

122. Ridley, *Discovery*, 48.

123. All information about Merian's journals comes from Sharon Valiant, "Maria Sibylla Merian: Recovering an Eighteenth Century Legend," *Eighteenth-Century Studies* 26, no. 3 (1993).

124. Qutd. in Schiebinger, *Plants and Empire*, 33-34.

125. Schiebinger, Plants and Empire, 32.

126. Maria Sibylla Merian, "Plate XLV," Metamorphosis insectorum Surinamensium (Tot Amsterdam, Voor den auteur..., als ook by Gerarde Valck, 1705). Translation from the Dutch by Patricia McNeill.

127. Sheibinger, Plants and Empire, 35.

128. Cheryl McEwan, "Gender, science, and physical geography in Nineteenth-Century Britain," *Area 30*, no. 3 (1998): 218-219.

129. John Kwadwo Osei-Tutu and Victoria Smith, "Interpreting West Africa's Forts and Castles" in *Shadows on Empire in West Africa*, eds. John Kwadwo Osei-Tutu and Victoria Smith (Palgrave Macmillan, 2018), 2.

130. Mary Orr, "The Stuff of Translation and Independent Female Scientific Authorship: The Case of Taxidermy…, anon. (1820)," *Journal of Literature and Science* 8, no. 1 (2015): 27.

131. Carl Thompson, "Women Travellers, Romantic Era Science and the Banksian Empire," *Notes and Records: The Royal Society Journal of the History of Science* (2019): 20.

132. "The Slavery Connection: Bexley Heritage Trust, 2007-2009." Antislavery Usable Past, 2009–2007. http://www.antislavery.ac.uk/items/show/23.

CHAPTER 7

133. Florence Fenwick Miller, *Harriet Martineau*, ed. John H. Ingram (London: W.H. Allen & Co., 1884), 221–222.

134. Geoffrey Cantor, et al., *Science in the Nineteenth-Century Periodical* (Cambridge University Press, 2004), 13.

135. Bernard Lightman, *Victorian Popularizers of Science: Designing Nature for New Audiences* (University of Chicago Press, 2007), 32.

136. Leigh and Rocke, *Chemistry in Regency England*, 31; Jane Marcet, *Conversations on Chemistry*, vol. 2 (London: Longman, Green, Longman, and Roberts, 1853).

137. Jeffery G. Leigh and Alan J. Rocke, "Women and Chemistry in Regency England: New Light on the Marcet Circle," *Ambix 63*, no. 1 (2016): 28.

138. Susan M. Lindee, "The American Career of Jane Marcet's Conversations on Chemistry," *Isis 82*, no. 1 (1991).

139. Qutd. in Lightman, *Victorian Popularizers*, 99.

140. Suzanne Le-May Sheffield, *Revealing New Worlds: Three Victorian Women Naturalists* (Routledge, 2013), 15.

141. Ibid., 16.

142. Ibid., 19.

143. Barbara T. Gates, *Kindred Nature: Victorian and Edwardian Women Embrace the Living World* (University of Chicago Press, 1998), 39.

144. Lightman, *Victorian Popularizers*, 155.

145. Alan Rauch, "Parables and Parodies: Margaret Garry's Audience in the Parables from Nature," *Children's Literature* 25 (1997): 141.

146. Jordan Larsen, "The Evolving Spirit: Morals and Mutualism in Arabella Buckley's Evolutionary Epic," *Notes and Records: Royal Society Journal of the History of Science* 71 (2017): 391–393.

147. Lightman, *Victorian Popularizers*, 239.

CHAPTER 8

148. Barbara T. Gates, *Kindred Nature: Victorian and Edwardian Women Embrace the Living World* (University of Chicago Press, 1998), 36; Merrill, Lynn L., *The Romance of Victorian Natural History* (New York: Oxford University Press, 1989), 8-9.

149. See Chapter 8.

150. Gates, *Kindred Nature*, 35-36.

151. Ann B. Shteir, *Cultivating Women, Cultivating Science: Flora's Daughters and Botany in England 1760-1860* (Johns Hopkins University Press, 1996), 35.

152. Meegan Kennedy, "Discriminating the 'Minuter Beauties of Nature': Botany as Natural Theology in a Victorian Medical School" in *Strange Science: Investigating the Limits of Knowledge in the Victorian Age* (University of Michigan Press, 2017), 44.

153. Londa Schiebinger, *Nature's Body: Gender in the Making of Modern Science* (Rutgers University Press, 1993), 17.

154. Kennedy, "Discriminating," 43.

155. Shteir, Cultivating Women, 64.

156. Ibid., 157.

157. Ibid., 157.

158. Ibid., 197.

159. Ibid., 225-226.

160. Tina Gianquitto, "Botanical Smuts and Hermaphrodites: Lydia Becker, Darwin's Botany, and Education Reform," *Isis* 104, no. 2 (June 2013): 250-251.

161. Shteir, *Cultivating Women*, 227.

162. Maureen Wright, *Elizabeth Wolstenholme Elmy and the Victorian Feminist Movement: The biography of an insurgent woman* (Manchester University Press, 2011), 1.

163. Leila McNeill, "The Early Feminist Who Used Botany To Teach Kids About Sex," The Atlantic, October 6, 2016, *https://www.theatlantic.com/science/archive/2016/10/the-early-feminist-who-used-botany-to-teach-kids-about-sex/503030/#:~:targetText=But%20when%20Elizabeth%20Wolstenholme%20Elmy,as%20a%20sex%2Deducation%20handbook.*

164. Gates, Kindred Nature, 132.

165. Elizabeth Wolstenholme Elmy, "Baby Buds" in *In Nature's Name: An Anthology of Women's Writing and Illustration 1780-1930*, ed. Barbara T. Gates (University of Chicago Press, 2002), 485.

CHAPTER 9

166. Susan M. Reverby, *Ordered to Care: The Dilemma of American Nursing, 1850-1945* (Cambridge: Cambridge University Press), 1987, 11.

167. Ibid., 12

168. Patricia D'Antonio, *American Nursing: A History of Knowledge, Authority, and the Meaning of Work* (Baltimore: The Johns Hopkins University Press, 2010), 5.

169. Ibid., 27.

170. Ibid., 13.

171. Ibid., 20. See also Charlotte Furth, *A Flourishing Yin: Gender in China's Medical History: 960-1665* (Berkeley: The University of California Press, 1999).

172. Althea T. Davis, *Early Black American Leaders in Nursing: Architects for Integration and Equality* (Jones and Bartlett Publishers and National League for Nursing, 1999): 3–4.

173. Ibid., 5. See also Katherine Bankole, *Slavery and Medicine: Enslavement and Medical Practices in Antebellum Louisiana* (New York, Garland Publishing, 1998).

174. Ibid., 7.

175. Ibid., 12.

176. "Classified Ad 29," *The New York Times*, September 25, 1861. ProQuest Historical Newspapers.

177. Ibid.

178. D'Antonio, American Nursing, 9.

179. Reverby, Ordered to Care, 47.

180. Florence Nightingale, *Notes on Nursing: What it is, and what it is not* (New York: D. Appleton and Company, 1860). A digitized, searchable transcription of this specific edition is available from the University of Pennsylvania's Digital Library: *https://digital.library.upenn.edu/women/nightingale/nursing/nursing.html*

181. Ibid.

182. D'Antonio, American Nursing, 6.

183. D'Antonio, American Nursing, 11.

184. Ibid., 16.

185. Ibid., 17.

186. Davis, *Early Black American Leaders*, 28–29.

187. D'Antonio, American Nursing, 26.

188. Davis, *Early Black American Leaders*, 32.

189. Ibid., 39.

190. Ibid., 40–41.

191. Ibid., 43.

192. Ibid., 46–48.

193. D'Antonio, American Nursing, 74.

194. Ibid., 26.

195. Marjorie N. Feld, Lillian Wald: *A Biography* (Chapel Hill: University of North Carolina Press, 2008), 33.

196. Ibid., 35.

197. Ibid., 8.

198. Elizabeth Free and Liping Bu, "The Origins of Public Health Nursing: The Henry Street Visiting Nurse Service," American Journal of Public Health 100, no. 7 (2010): 1206–1207.

199. *https://socialwelfare.library.vcu.edu/people/wald-lillian/*

200. Lillian Wald, *The House on Henry Street* (New York: Henry Holt and Company, 1912), 29.

201. Ibid., 32.

202. Feld, Lillian Wald, 35.

CHAPTER 10

203. See Thomas Neville Bonner, *To The Ends of the Earth: Women's Search for Education in Medicine* (Cambridge: Harvard University Press, 1995).

204. Leila McNeill, "Dr. Anna Fischer-Dückelmann as Naturopath and Physician for Women in Imperial Germany," (Masters Thesis, University of Oklahoma, 2014), 53.

205. Ibid., 18–20.

206. Ibid., 33.

207. Ibid., 27.

208. Ibid., 34.

209. Ibid.

210. Ibid., 2.

211. Ibid., 55.

212. Debra Michals, "Elizabeth Blackwell," National Women's History Museum, online. *https://www.womenshistory.org/education-resources/biographies/elizabeth-blackwell Last accessed September 9, 2019.*

213. Regina Markell Morantz-Sanchez, *Sympathy and Science: Women Physicians in American Medicine* (Oxford: Oxford University Press, 1985), 48.

214. Ibid., 49.

215. Ibid., 65.

216. Ibid., 76

217. Sarah Ross Pripas-Kapit, "Educating Women Physicians of the World: International Students of the Women's Medical College of Pennsylvania, 1883-1911" (PhD Diss., University of California, Los Angeles, 2015), 1.

218. Ibid., 5.

219. Sarah Ross Pripas-Kapit, "Educating Women Physicians of the World: International Students of the Women's Medical College of Pennsylvania, 1883-1911" (PhD Diss., University of California, Los Angeles, 2015), 4.

220. Leila McNeill, "This 19th Century 'Lady Doctor' Helped Usher Indian Women Into Medicine," *Smithsonian Magazine* August 24, 2017. Online *https://www.smithsonianmag.com/science-nature/19th-century-lady-doctor-ushered-indian-women-medicine-180964613/ Last accessed September 9, 2019.*

221. Josambi, "Anandibai Joshee," 3190.

222. McNeill, "19th Century 'Lady Doctor.'"

223. Pripas-Kapit, "Educating Women Physicians," 51.

224. Ibid., 55–56.

225. Ibid., 57.

226. Ibid., 57–58.

227. Sarah Pripas-Kapit, "'We Have Lived on Broken Promises': Charles A. Eastman, Susan La Flesche Picotte, and the Politics of American Indian Assimilation during the Progressive Era," *Great Plains Quarterly* 35, no.1 (2015), 54–55.

228. Ibid., 55.

229. Pripas-Kapit, "Educating Women Physicians," 78–79.

230. Ibid., 80.

231. Valerie Sherer Mathes, "Susan La Flesche Picotte, M.D.: Nineteenth-Century Physician and Reformer," *Great Plains Quarterly* 13, no. 3 (1993), 178.

232. Quoted in Valerie Sherer Mathes, "Susan La Flesche Picotte, M.D," 174.

233. Pripas-Kapit "Educating Women Physicians", 92.

234. Ibid., 97.

235. Ibid., 100.

236. Ibid., 101.

CHAPTER 11

237. Meera Josambi, "Anandibai Joshee: Retrieving a Fragmented Feminist Image," *Economic and Political Weekly* 31, no 49 (1996), 3192.

238. Dorothea Klumpke, "Women's Work in Astronomy" in *The Observatory* (1899), 299–300.

239. See Chapter 4.

240. David Alan Grier, When Computers Were Human, (Princeton University Press, 2005), 5.

241. See Chapter 4.

242. Grier, *When Computers Were Human*, 20-25.

243. Margaret W. Rossiter, "'Women's Work' in Science, 1880-1910," Isis 71, no. 3 (1980): 382.

244. Qutd. in Grier, *When Computers were Human*, 83.

245. Qutd. in Sobel, *The Glass Universe: How The Ladies of the Harvard Observatory Took the Measure of the Stars* (NY: Viking University Press, 2016), 96.

246. Pamela Mack, "Strategies and Compromises: Women in Astronomy at Harvard College Observatory, 1870-1929," *Journal for the History of Astronomy* 21, no. 1 (1990): 70.

247. Qutd. in Mack, "Strategies and Compromises," 70.

248. Mary Brück, "Slave-Wage Earners" in *Women in Early British and Irish Astronomy* (Springer, 2009), 203.

249. Peggy Aldrich Kidwell, "Women Astronomers in Britain, 1780-1930," Isis 75, no. 3 (1984): 536.

250. Brück, "Slave-Wage Earners," 203.

251. Ogilvie, "Obligatory amateurs: Annie Maunder (1868-1947) and British women astronomers at the dawn of professional astronomy," *British Journal for the History of Science* 33 (2000): 73.

252. M.T. Brück, "Lady Computers at Greenwich in the early 1890s," *Quarterly Journal of the Royal Astronomical Society* 36 (1995): 90.

253. Brück, "Slave-Wage Earners," 204.

254. T. Stevenson, "Making Visible the First Women in Astronomy in Australia: The Measurers and Computers

Employed for the Astrographic Catalogue," *Publications of the Astronomical Society of Australia* 31 (2014): 2.

255. Ibid., 2–3.

256. Brück, "Slave-Wage Earners," 216.

257. Stevenson, "Making Visible," 5.

258. Ibid., 7.

259. Ibid., 4.

260. D. Hoffleit, "Women in the History of Variable Star Astronomy," *American Association of Variable Star Observers* (1993): 6.

CHAPTER 12

261. Harriet Gillespie, "Labor-Saving Devices Supplant Servants," Good Housekeeping, January 1913, 132-134.

262. Ibid., 133.

263. Elisa Miller, "In the Name of the Home: Women, Domestic Science, and American Higher Education, 1864-1930" (PhD Diss., University of Illinois Urbana-Champaign, 2003), 10–11.

264. Laurel D. Graham, "Domesticating Efficiency: Lillian Gilbreth's Scientific Management of Homemakers, 1924-1930," *Signs* 24, no. 3 (1999), 646.

265. Ibid., 52-53.

266. "Houseworkers and New Apparatus," *Good Housekeeping*, January 1913, 135–136.

267. Graham, "Domesticating Efficiency," 637.

268. Ibid., 638.

269. Ibid., 638-639.

270. Ibid., 639.

271. Ibid.

272. Ibid., 642.

273. Ibid., 659.

274. Ibid.

275. Leila McNeill, "The History of Breeding Mice for Science Begins With a Woman in a Barn," Smithsonian.com, March 20, 2018, *https://smithsonianmag.com/science-nature/history-breeding-mice-science-leads-back-woman-barn-180968441/ Last accessed September 18, 2019.*

276. Ibid.

277. Karen Rader, *Making Mice: Standardizing Animals for American Biomedical Research, 1900-1955* (Princeton: Princeton University Press, 2004), 42.

278. David P. Steensma, et. al., "Abbie Lathrop, the "Mouse Woman of Granby": Rodent Fancier and Accidental Genetics Pioneer," *Mayo Clinic Proceedings* 85, no. 11 (2010), e83.

279. [Brooklyn Eagle] "Woman Runs a Mouse Farm," *The Washington Post* June 20, 1909, M4.

280. [New York Press] "Raises Rats and Mice." *The Los Angeles Times* December 26, 1907, 16.

CHAPTER 13

281. Hall, Lesley A, "Stopes [married name Roe], Marie

Charlotte Carmichael (1880–1958), sexologist and advocate of birth control," *Oxford Dictionary of National Biography*, 2004. Accessed September 24, 2019. *https://www-oxforddnb-com.ezproxy.lib.ou.edu/view/10.1093/ref:odnb/9780198614128.001.0001/odnb-9780198614128-e-36323.*

282. Laura Doan, "Marie Stopes's Wonderful Rhythm Charts: Normalizing the Natural," *Journal of the History of Ideas*, 78, no. 4 (2017), 595-620.

283. Wendy Kline, *Building a Better Race: Gender, Sexuality, and Eugenics from the Turn of the Century to the Baby Boom* (Berkeley, University of California Press, 2005): 13.

284. Ibid., 13, 20.

285. Greta Jones, "Women and eugenics in Britain: The case of Mary Scharlieb, Elizabeth Sloan Chesser, and Stella Browne," *Annals of Science* 52:5 (1995): 485.

286. Ibid., 489.

287. "Notes," Nature March 17, 1921, 88. *https://books.google.co.uk/books?id=3-4RAAAAYAAJ&pg*

288. Esther Katz, "Sanger, Margaret (14 September 1879–06 September 1966), birth control advocate." *American National Biography*. 1 Feb. 2000; Accessed 24 Sep. 2019. *https://www-anb-org.ezproxy.lib.ou.edu/view/10.1093/anb/9780198606697.001.0001/anb-9780198606697-e-1500598.*

289. For more on visiting nurses and settlement houses, see Chapter 9, this volume.

290. Margaret Sanger, ""The Woman Rebel" and The Fight for Birth Control," [Apr 1916] .Published article. Source: Records of the Joint Legislative Committee to Investigate Seditious Activities, New York State Archives , Margaret Sanger Microfilm, C16:1035. *https://www.nyu.edu/projects/sanger/webedition/app/documents/show.php?sangerDoc=306320.xml*

291. "No Gods, No Masters": Margaret Sanger on Birth Control," History Matters, George Mason University. Accessed December 26, 2019.http://historymatters.gmu.edu/d/5084/.

292. Debra Michals, "Margaret Sanger." National Women's History Museum, 2017. Accessed September 27, 2019. www.womenshistory.org/education-resources/biographies/margaret-sanger.

293. Cathy Moran Hajo, *Birth Control on Main Street: Organizing Clinics in the United States, 1916-1939* (University of Illinois Press, 2010): 11.

294. Jennifer Young, "An Emancipation Proclamation to the Motherhood of America," *Lady Science*, November 16, 2017. *https://www.ladyscience.com/emancipation-proclamation-to-the-motherhood-of-america/no38.*

295. Ibid., 12-13.

296. Cathy Moran Hajo, "What Every Woman Should Know: Birth Control Clinics in the United States, 1916-1940," PhD Diss. (New York University, 2006): 321-322.

297. Ibid., 323.

298. Ibid., 327.

299. Ibid., 327.

300. Lakshmeeramya Malladi, "United States v. One Package of Japanese Pessaries (1936)". *Embryo Project Encyclopedia* (2017-05-24). ISSN: 1940-5030 http://embryo.asu.edu/handle/10776/11516.

CHAPTER 14

301. Zelia Nuttall, "New Year of Tropical American Indigenes: The New Year Festival of the Ancient Inhabitants of Tropical America and Its Revival," *Pan American Miscellany*, no. 9 (1928): 7.

302. Carmen Ruiz, "Insiders and Outsiders in Mexican Archaeology (1890-1930)" (PhD diss., University of Texas at Austin, 2003), 32.

303. Zelia Nuttall, "Ancient Mexican Superstitions," *The Journal of American Folklore* 10, no. 39 (1897): 265-266.

304. David L. Brownman, *Cultural Negotiations: The Role of Women in the Founding of Americanist Archaeology* (University of Nebraska Press, 2013), 5.

305. Annual report of the trustees of the Peabody Museum of American Archaeology and Ethnology, vol. III (John Wilson and Son, 1887), 566.

306. Ruiz, "Insiders and Outsiders," 36.

307. Nuttal, "Ancient Mexican Superstitions," 265.

308. Ibid., 281.

309. Qtd. in Ruiz, "Insiders and Outsiders," 249.

310. Margaret M. Bruchac, *Savage Kin: Indigenous Informants and American Anthropologists* (University of Arizona Press, 2018), 33.

311. Ibid., 29.

312. Ibid., 86 & 83.

313. Ibid., 108.

314. Ibid., 109.

315. Qutd. in Bruchac, 110.

316. Brownman, Cultural Negotiations, 128.

317. M.R. Harrington, "Man and the Beast in Gypsum Cave," *Desert Magazine* 3, no. 6, April 1940, 5.

318. M.R. Harrington, "Ashes Found With Sloth Remains," *The Science News-Letter* 17, no. 478, (June 17, 1930): 365.

319. Ibid., 103.

320. Ibid., 127.

321. Ibid., 130.

CHAPTER 15

322. "The S-1 Committee," Atomic Heritage Foundation, April 27, 2017, *https://www.atomicheritage.org/history/s-1-committee.*

323. Qutd. in Ruth H. Howes and Caroline L. Herzenberg, *Their Day in the Sun: Women of the Manhattan Project,* (Philadelphia: Temple University Press, 1999), 39.

324. Ibid., 14

325. Ibid., 98. See also Chapter 12.

326. Jennifer Light, "When Computers Were Women,"

Technology and Culture 20, no. 3. (1999): 471-472.

327. Ibid., 13.

328. Leon Lidofsky, "Chien-Shiung Wu, 29 May 1912. 16 February 1997," *Proceedings of the American Philosophical Society* 145, no.1 (2001): 119.

329. Leslie R. Groves, *Now It Can Be Told: The Story of the Manhattan Project* (Harper & Brothers, 1962), 166.

330. Howes and Herzenberg, *Their Day in the Sun,* 16.

331. Denise Kiernan, *The Girls of Atomic City: The Untold Story of the Women who Helped Win World War II* (NY: Touchstone, 2013), 86.

332. Toshihiro Higuchi, "Epistemic frictions: radioactive fallout, health risk assessments, and the Eisenhower administration's nuclear-test ban policy, 1954-1958," *International Relations of the Asia-Pacific* 18 (2018): 100.

333. Ibid., 101.

334. Simikoh, "A Life Story of Saruhashi Katsuko (1929-2007)," Contemporary Japaense Feminist Debates at Penn, 2016, *https://japanfeministdebates.wordpress.com/2016/11/30/a-life-story-of-saruhashi-katsuko-1920-2007/.*

335. Ibid.

CHAPTER 16

336. Qinna Shen, "A Refugee Scholar from Nazi Germany: Emmy Noether and Bryn Mawr College," *The Mathematical Intelligencer* 41, no. 3 (2019): 53.

337. Ibid., 65.

338. Ibid., 55.

339. Qutd. in Shen, "A Refugee Scholar," 55.

340. Ibid., 55.

341. Ibid., 60.

342. Qutd. in Michael Cavna, "Emmy Noether Google Doodle: Why Einstein called her a 'creative mathematical genius,'" *The Washington Post*, March 23, 2015, *https://www.washingtonpost.com/news/comic-riffs/wp/2015/03/23/emmy-noether-google-doodle-why-einstein-called-her-a-creative-mathematical-genius/.*

343. Evelyn Lamb, "Hermann Weyl's Poignant Eulogy for Emmy Noether," *Scientific American*, November 23, 2016, *https://blogs.scientificamerican.com/roots-of-unity/hermann-weyls-poignant-eulogy-for-emmy-noether/*

344. Shen, "A Refugee Scholar," 65.

345. Arnold Reisman, "Hilda Geiringer: A Pioneer of Applied Mathematics and a Woman Ahead of Her Time Was Saved from Fascism by Turkey," *Women in Judaism: A Multidisciplinary Journal* 4, no. 2 (Fall 2007): 2.

346. Alp Eden and GürolIrzik, "German mathematician in exile in Turkey: Richard von Mises, William Prager, Hilda Geiringer, and their impact on Turkish mathematics," *Historia Mathematica* 39 (2012): 442.

347. Qutd. in Siegmund-Schultze, *Mathematicians Fleeing from Nazi Germany: Individual Fates and Global Impact* (Princeton University Press, 2009), 369-370.

Afterword

348. Qutd. in Margaet W. Rossiter, *Women Scientists in America: Before Affirmative Action*, 1940-1972 (Johns Hopkins University Press, 1995), 36.

349. Henry Lang, "Tilly Edinger's Deafness," *Tilly Edinger: Leben Und Werk Einer Jüdischen Wissenschaftlerin vol. 1*, eds. Rolf Kohring and Gerald Kreft (Schweizerbart Sche Vlgsb., 2003), 362.

350. Ibid., 362

351. Alice Hamilton, *Exploring the Dangerous Trades: The Autobiography* (Little, Brown and Company, 1943), 397-398.

352. Phyllis Appel, "Dr. Tilly Edinger, 1897-1967, Paleoneurologist" in *The Jewish Connection: Profiles of the Famous and Infamous* (Graystone Enterprises LLC, 2013).

353. Emily A. Bucholtz and Ernst-August Seyfarth, "The gospel of the fossil brain: Tilly Edinger and the science of paleoneurology," *Brain Research Bulletin* 48, no. 4 (199): 356.

354. Lang, "Tilly Edinger's Deafness," 367.

367. Ibid., 14.

368. Unger, *Beyond Nature's Housekeepers*, 87.

369. Qutd. in Merchant, "Women of the Progresive Conservation Movement," 59.

370. Jack E. Davis, "'Conservation Is Now a Dead Word': Marjory Stoneman Douglas and the Transformation of American Environmentalism," *Environmental History* 8, no. 1 (2003): 56

371. Qutd. in Mary Anne Peine, "Women for the Wild: Douglas, Edge, Murie and the American conservation movement." MA thesis, (University of Montana, 2002), 18.

372. Qutd. in Davis, "'Conservation,'" 59.

373. Marjory Stoneman Douglas, *The Everglades: River of Grass* (Rinehart & Company, Inc., 1947), 5.

374. Davis, "'Conservation,'" 62.

375. Unger, *Beyond Nature's Housekeepers*, 95-97.

376. Ibid., 99.

CHAPTER 17

355. Qutd. in Eileen Boris, "The Power of Motherhood: Black and White Activist Women Redefine the "Political'," *Yale Journal of Law and Feminism* 2, no. 25 (1989): 25.

356. Qutd. in Nancy C. Unger, *Beyond Nature's Housekeepers: American Women in Environmental History* (NY: Oxford University Press, 2012), 84-85.

357. Qutd. in Carolyn Merchant, "Women of the Progressive Conservation Movement: 1900-1916," *Environmental Review* 8, no. 1 (1984): 74-75.

358. See Chapter 13.

359. Leila McNeill, "The First Female Student at MIT Started An All-Women Chemistry Lab and Fought for Food Safety," Smithsonian.com, December 18, 2018, *https://www.smithsonianmag.com/science-nature/first-female-student-mit-started-women-chemistry-lab-food-safety-180971056/*.

360. Robert K. Musil, *Rachel Carson and Her Sisters, Extraordinary Women Who Have Shaped America's Environment* (Rutgers University Press, 2014), 48.

361. Babara Haber, "Cooking" in *The Reader's Companion to U.S. Women's History*, eds. Wilma Mankiller, Gwendolyn Mink, Marysa Navarro, Barbara Smith, and Gloria Steinam (NY: Houghton Mifflin Company, 1998), 136.

362. Ellen Henrietta Richards, *The Art of Right Living* (Whitcomb & Barrows, 1904), 47.

363. Boris, "The Power of Motherhood," 32.

364. Ibid., 27.

365. Rebecca Cole, "First Meeting of the Women's Missionary Society of Philadelphia," *The Women's Era* 3, no. 4 (1896): 5.

366. Mary Church Terrell, "The progress of colored women," Address delivered to the National Woman's Suffrage Association, Columbia Theatre, Washington, D.C., 1898, 10-11.

CHAPTER 18

377. Mamie Phipps Clark, "Interview of Dr. Mamie Clark by Ed Edwin, May 25, 1976." Interview by Ed Edwin in *The Reminiscences of Mamie Clark* (Alexandria: Alexander Street Press, 2003), 101.

378. Ibid.

379. Axelle Karera (2010) and Alexandra Rutherford (2017), "Profile of Mamie Phipps Clark," in Psychology's Feminist Voices Multimedia Internet Archive, ed. A. Rutherford. Accessed December 23, 2019, http://www.feministvoices.com/mamie-phipps-clark/.

380. Clark, "Interview of Dr. Mamie Clark."

381. Clark, "Interview of Dr. Mamie Clark."

382. Karera, "Profile of Mamie Phipps Clark."

383. Clark, "Interview of Dr. Mamie Clark."

384. Karera, "Profile of Mamie Phipps Clark."

385. "The Significant of the 'Doll Test,'" NAACP Legal Defense Fund, Accessed November 29, 2019, *https://www.naacpldf.org/ldf-celebrates-60th-anniversary-brown-v-board-education/significance-doll-test/*

386. Shirley Mahaley Malcom, Paula Quick Hall, Janet Welsh Brown, "The Double Bind: The Price of Being a Minority Woman in Science," Conference Proceedings, American Association for the Advancement of Science (76-R-3) 1976), ix.

387. Ibid., 1.

388. Ibid., 8–9.

389. Ibid., 18.

390. Ibid., 35.

391. Olivia A. Scriven, "The Politics of Particularism: HBCUs, Spelman College, and the Struggle to Educate Black Women in Science, 1960–1997." (PhD Diss, Georgia Institute of Technology, 2006), 10.

Endnotes

392. Ibid., 77.

393. Ibid., 16.

394. Ibid., 93.

395. Ibid, 86-87.

396. Ibid., 130.

397. May Edwin Mann Burke, "The Contributions of Flemmie Pansy Kittrell to Education Through Her Doctrines on Home Economics," (PhD Diss University of Maryland, 1988), 6–11.

398. Ibid., 12.

399. Allison Beth Horrocks, "Good Will Ambassador with a Cookbook: Flemmie Kittrell and the International Politics of Home Economics," (PhD Diss, University of Connecticut, 2016), 2.

400. Ibid., 14.

401. Quoted in Horrocks, "Good Will Ambassador with a Cookbook," 46.

402. Burke, "The Contributions of Flemmie Pansy Kittrell," 24. For more on the place of home economics and domestic engineering in American society, see Chapter 13, this volume.

403. Ibid., 29.

404. Horrocks, "Good Will Ambassador with a Cookbook," 7.

405. Burke, "The Contributions of Flemmie Pansy Kittrell," 37.

406. Ibid., 36.

407. Ibid., 34–35.

408. Horrocks, "Good Will Ambassador with a Cookbook," 9.

CHAPTER 19

409. "Valentina Tereshkova," Smithsonian National Air and Space Museum, Accessed October 24, 2019. https://airandspace.si.edu/people/historical-figure/valentina-tereshkova

410. Asif A. Siddiqi, Sputnik and the Soviet Space Challenge (Gainesville, University Press of Florida, 2003), 353.

411. Smithsonian, "Valentina Tereshkova,".

412. Ibid.

413. Barbara Evans Clements, "Tereshkova, Valentina," in The Oxford Encyclopedia of Women in World History, ed. Bonnie G. Smith (Oxford: Oxford University Press, 2008).

414. Quoted in Siddiqi, 354.

415. Spaceport News June 20, 1963. All issues of Spaceport News cited in this volume are from the Kennedy Space Center Files, Office of Manned Space Flight, Public Affairs Office, National Archives and Records Administration, Atlanta, GA.

416. Ibid., 2.

417. See Chapter 12, this volume.

418. See also Margot Lee Shetterly, Hidden Figures: The American Dream and the Untold Story of the Black Women Who Helped Win the Space Race (New York: William Morrow & Company, 2016).

419. Matthew Sanders, "The Woman Who Got Real Food to Space," Smithsonian National Air and Space Museum, April 9, 2018. https://airandspace.si.edu/stories/editorial/woman-who-got-real-food-space.

420. Ibid.

421. Quoted in Anna Reser, "The Lost Stories of NASA's 'Pink-Collar' Workforce, The Atlantic February 15, 2017. https://www.theatlantic.com/science/archive/2017/02/ursula-vils-nasa/516468/. See also Anna Reser, "Images of Place in American Spaceflight, 1958-1974." (PhD diss, University of Oklahoma, 2019).

422. Sanders

423. "NASA Johnson Space Center Oral History Project Biographical Data Sheet, Dee O'Hara." National Aeronautics and Space Administration, Johnson Space Center Oral History Project. https://historycollection.jsc.nasa.gov/JSCHistoryPortal/history/oral_histories/OHaraDB/OHaraDB_Bio.pdf.

424. Dee O'Hara, NASA Johnson Space Center Oral History Project Oral History Transcript, Dee O'Hara," interviewed by Rebecca Wright, Mountain View California, April 23, 2002. https://historycollection.jsc.nasa.gov/JSCHistoryPortal/history/oral_histories/OHaraDB/OHaraDB_4-23-02.pdf.

425. Ibid., 33.

426. For the women who underwent unofficial medical testing in the 1960s and advocated for women astronauts, see Margaret Weitekamp, Right Stuff Wrong Sex: America's First Women in Space Program (Baltimore: The Johns Hopkins University Press, 2004).

427. Quoted in Amy E. Foster, Integrating Women into the Astronaut Corps: Politics and Logistics at NASA, 1972-2004 (Baltimore: The Johns Hopkins University Press, 2011), 88.

428. Ibid., 95.

429. Ibid., 99.

430. Ibid.

431. Ibid., 100.

432. Ibid., 101

433. Ibid., 114-115.

434. NASA Biographical Data, "Mae C. Jemison," National Aeronautics and Space Administration (nd): https://www.nasa.gov/sites/default/files/atoms/files/jemison_mae.pdf.

435. NASA Biographical Data, "Chiaki Muka," National Aeronautics and Space Administration (nd): https://www.nasa.gov/sites/default/files/atoms/files/mukai.pdf.

CHAPTER 20

436. Clare Boothe Luce, "But Some People Simply Never Get the Message," Life June 28, 1963: 31.

437. Chapter 21

438. Robert H. Lowie and Leta Stetter Hollingworth, "Science and Feminism," The Scientific Monthly 3,

Afterword

no. 3 (1916): 277.

439. Jill G. Morawski and Gail Agronick, "A Restive Legacy: The History of Feminist Work in Experimental and Cognitive Psychology," *Psychology of Women Quarterly* 15 (1991): 570.

440. James Capshew and Alejandra C. Laszlo, "'We would not take no for an answer': Women Psychologists and Gender Politics During World War II," *Journal of Social Issues* 42, no. 1 (1986): 160-162.

441. Stephanie A. Shields, "Ms. Pilgrim's Progress: The Contributions of Leta Stetter Hollingworth to the Psychology of Women," *American Psychologist* 30, no. 8 (1975): 853.

442. Ibid., 854.

443. Alexandra Rutherford and Leeat Granek, "Emergence and Development of the Psychology of Women" in *Handbook of Gender Research in Psychology*, eds. J.C. Chrisler and D.R. McCreary (Springer Science+Business Media, LLC, 2010), 19.

444. Lisa Held, "Leta Hollingworth," *Psychology's Feminist Voices*, 2010, http://www.feministvoices.com/leta-hollingworth/

445. Edward Lee Thorndike, *Educational Psychology: Mental work and fatigue and individual differences and their causes* (Teacher's College, Columbia University, 1921), 188.

446. Leta Stetter Hollingworth, "Variability as Related to Sex Differences in Acheivement: A Critique," *American Journal of Sociology* 19, no. 4 (1914): 526.

447. Lowie and Hollingworth, "Science and Feminism," 283.

448. Qutd. in Capshew and Laszlo, "'We would not take no,'" 163.

449. Rutherford and Granek, "Emergence and Development," 24.

450. Amy Johnson and Elizabeth Jonston, "Unfamiliar Feminisms: Revisiting the National Council of Women Psychologists," *Psychology of Women Quarterly* 34 (2010): 311.

451. See Chapter 19 for more on Clark.

452. Naomi Weisstein, "'How can a little girl like you teach a great big class of men?' the Chairman Said, and Other Adventures of a Woman in Science" in *Working It Out: 23 Women Writers, Artists, Scientists, and Scholars Talk About Their Lives and Work*, eds. Sara Ruddick and Pamela Daniels (NY: Pantheon Books, 1977), 243.

453. Ibid., 244.

454. Naomi Weisstein, "Psychology Constructs the Female; or, The Fantasy Life of the Male Psychologist (with Some Attention to the Fantasies of His Friends, the Male Biologist and the Male Anthropologist)," *Feminism and Psychology* 3, no. 2 (1993): 195

455. Alexandra Rutherford, Kelli Vaughn-Blount, and Laura C. Ball, "Responsible Opposition, Disruptive Voices: Science, Social Change, and the History of Feminist Psychology," *Psychology of Women Quarterly* 34 (2010): 464.

456. Qutd. in Martha T. Mednick and Laura L. Urbanski, "The Origins and Activities of APA's Division of the Psychology of Women," *Psychology of Women Quarterly* 15 (1991): 651.

CHAPTER 21

457. Ibid., 655.

458. Lowie and Hollingworth, "Science and Feminism," 284.

459. "Coeducation: History of Women at Princeton University," Princeton University, Accessed November 29, 2019, *https://libguides.princeton.edu/c.php?g=84581&p=543232.*

460. Vera C. Rubin, "An Interesting Voyage," *The Annual Review of Astronomy and Astrophysics* 49 (2011), 3.

461. Kristine Larsen, "Reminiscences on the Career of Martha Stahr Carpenter" Between a Rock and (Several) Hard Places," *Journal of the American Association of Variable Star Observers* 40 (2012), 55.

462. "Young Mother Figures Center of Creation by Star Motions," *The Washington Post* December 31, 1950. ProQuest.

463. Rubin, "An Interesting Voyage," 4.

464. Rubin, "An Interesting Voyage," 9.

465. "A History of Palomar Observatory, Caltech, Accessed November 29, 2019, http://www.astro.caltech.edu/palomar/about/history.html#55

466. Rubin, "An Interesting Voyage," 12.

467. Ibid., 13. See Vera Rubin, et. al., "Rotation of the Andromeda Nebula from a Spectroscopic Survey of Emission Regions" *Astrophysics Journal*, 159 (1970), 379.

468. Robert Lambourne, "Interview with Jocelyn Bell Burnell," *Physics Education* 3, no. 183 (1996),183-186.

469. Jocelyn Bell Burnell, "Petit Four," *Annals of the New York Academy of Sciences* 302, no. 1 (1977), 685.

470. Ibid, 685–686.

471. Ibid.

472. Ibid., 687.

473. Antony Hewish, Jocelyn Bell Burnell, et. al., "Observation of a Rapidly Pulsating Radio Source," *Nature* 217 (1968), 709–713.

474. "The Nobel Prize in Physics 1974," Accessed December 23, 2019. *https://www.nobelprize.org/prizes/physics/1974/summary/*

475. Jocelyn Bell Burnell, "So Few Pulsars, So Few Females," *Science* 304, no. 5670 (2004), 489.

476. Ibid.

477. "The Nobel Medal for Physics and Chemistry," The Nobel Prize. Accessed November 29, 2019,

478. *https://www.nobelprize.org/prizes/facts/the-nobel-medal-for-physics-and-chemistry-2.*

Bibliography

INTRODUCTION

Gaida, Margaret. "Muslim Women and Science: The Search for the "Missing" Actors," *Early Modern Women* vol. 11, no. 1 (Fall 2016), 197–206

Maisels, Charles Keith. *The Emergence of Civilization.* London: Routledge, 1990.

Meador,Betty De Shong. *Inanna, lady of largest heart: poems of the Sumerian high priestess Enheduanna.* Austin: University of Texas Press, 2000.

Ogilvie, Marilyn and Joy Harvey, eds. *The Biographical Dictionary of Women in Science, Volume 1.* New York: Routledge, 2000.

CHAPTER 1

French, Valerie. "Midwives and Maternity Care in the Roman World." in *Midwifery and the Medicalization of Childbirth: Comparative Perspectives,* ed. Edwin van Teijlingen, George Lowis, Peter McCaffery, and Maureen Porter (New York: Nova Science Publishers, 2004), 543–62.

Furth, Charlotte. *A Flourishing Yin: Gender in China's Medical History: 960-1665.* Berkely: University of California Press, 1999.

Hanson, Ann Ellis. "'Diseases of Women 1.'" *Signs* 1:2 (1975), 576–584.

Nunn, J. F. *Ancient Egyptian Medicine.* Norman: University of Oklahoma Press, 1996.

Robbins, Robbins. *Women in Ancient Egypt.* London: British Museum Press, 1993.

Siraisi, Nancy G. *Medieval and Early Renaissance Medicine: An Introduction to Knowledge and Practice.* Chicago: The University of Chicago Press, 1990.

CHAPTER 1

Bicknell, Peter. "The Witch Aglaonice and Dark Lunar Eclipses in the Second and First Centuries BC." *Journal of the British Astronomical Association,* no. 93, 1983, 160–63.

Goldstein, Bernard R. Review of *Babylonian Eclipse Observations from 750 BC to 1 BC,* edited by Peter J. Huber and Salvo De Meis. *Aestimatio,* vol. 1, 2001.

Hill, D. E. "The Thessalian Trick." *Rheinisches Museum Für Philologie,* vol. 116, 1973, 221–38.

Hollywood, Amy. "'Who Does She Think She Is?':

Christian Women's Mysticism." *Theology Today,* no. 60, 2003, 5–15.

Jones, Sharon, and Diana Neal. "Negotiable Currencies: Hildegard of Bingen, Mysticism and the Vagaries of the Theoretical." *Feminist Theology,* vol. 11, no. 3, 2003, 375–84.

Maddocks, Fiona. *Hildegard of Bingen: The Woman of Her Age.* Doubleday, 2001.

Newman, Marsha. "Christian Cosmology in Hildegard of Bingen's Illuminations." *A Journal of Catholic Thought and Culture,* vol. 5, no. 1, 2002, 41–61.

Newman, Barbara. *Sister of Wisdom: St. Hildegard's Theology of the Feminine.* University of California Press, 1987.

Ovid. *Metamorphoses.* 4th ed., B. Law, 1797.

Phillips, Oliver. "The Witches' Thessaly." *Magic and Ritual in the Ancient World,* edited by Paul Mirecki and Marvin Meyer, vol. 141, Brill, 2002, 378–306.

Plato. *Gorgias.* Translated by Benjamin Jowett, 380AD, *https://ebooks.adelaide.edu.au/p/plato/p71g/complete. html.*

Plutarch. "Conjugalia Praecepta." *Moralia.* Translated by Frank Cole Babbitt, Harvard.

University Press, 1928.

Steele, J., et al. "The Accuracy of Eclipse Times Measured by the Babylonians." *Journal for the History of Astronomy,* 1997, 337–45.

Stothers, Richard B. "Dark Lunar Eclipses in Classical Antiquity." *Journal of the British Astronomical Association,* 1986, 95–97.

CHAPTER 3

Agnesi, Maria Gaetana. *Analytical Institutions in Four Books: Originally Written in Italian.* Translated by Rev. John Colson, Taylor and Wilks, London, 1801.

Bennett Peterson, Barbara, editor. *Notable Women of China: Shang Dynasty to the Early Twentieth Century.* Routledge, 2000.

Bernardi, G. "Nicole-Reine Étable de La Brière Lepaute (1723–1788)." *The Unforgotten Sisters,* Springer International Publishing, 2016, *https://link-springer-com. ezproxy.lib.ou.edu/chapter/10.1007/978-3-319-26127-0_19.*

Elman, Benjamin A. *A Cultural History of Modern Science in China.* Harvard University Press, 2006.

Findlen, Paula. "Calculations of Faith: Mathematics, Philosophy, and Sanctity in 18th-Century Italy (New Work on Maria Gaetana Agnesi)." *Historia Mathematica*, vol. 38, no. 2, 2010, 248–91.

Grier, David Alan. *When Computers Were Human.* Princeton University Press, 2005.

Ki Che Leung, Angela. "Wang Zhenyi." *Biographical Dictionary of Chinese Women*, edited by Lily Xiao Hong Lee et al., translated by W. Zhang, vol. 1: The Qing Period, 1644–1911, Routledge, 1998, 230–32.

Mazzotti, Massimo. *The World of Maria Gaetana Agnesi, Mathematician of God.* Johns Hopkins University Press, 2007.

Mazzotti, Massimo."Maria Gaetana Agnesi: Mathematics and the Making of the Catholic Enlightenment." *Isis*, vol. 92, no. 4, 2001, 657–83.

Roberts, Meghan K. "Learned and Loving: Representing Women Astronomers in Enlightenment France." *Journal of Women's History*, vol. 29, no. 1, 2017, 14–37.

Schiebinger, Londa. *The Mind Has No Sex?: Women in the Origins of Modern Medicine.* Harvard University Press, 1989.

Swerdlow, N. M. "Urania Propitia, Tabulae Rudophinae Faciles Redditae a Maria Cunitia Beneficent Urania, the Adaption of the Rudolphine Tables by Maria Cunitz." *A Master of Science History: Essays in Honor of Charles Coulston Gillispie*, vol. 30, Springer, 2012, 81–121.

Whaley, Leigh A. *Women's History as Scientists: A Guide to the Debates.* ABC-CLIO, 2003.

CHAPTER 4

Ashworth Jr., William B. "Scientist of the Day – Elisabeth Hevelius." *Linda Hall Library*, 2017, https://www.lindahall.org/elisabeth-hevelius/.

Bernardi, Gabriella. "Elisabetha Catherina Koopman Hevelius (1647-1693)." *The Unforgotten Sisters*, Springer International Publishing, 2016, 67–74.

Cook, Alan. "Ladies in the Scientific Revolution." *Notes and Records: The Royal Society Journal of the History of Science*, vol. 51, no. 1, 1997, 1–12.

DiMeo, Michelle. "'Such a Sister Became Such a Bother': Lady Ranelagh's Influence on Robert Boyle." *Intellectual History Review*, vol. 25, no. 1, 2015, 21–36.

Fara, Patricia. *Pandora's Breeches: Women, Science & Power in the Enlightenment.* Pimlico, 2004.

Herschel, Caroline. *Memoir and Correspondence of Caroline Herschel.* Edited by Mrs. John Herschel, 2nd ed., John Murray, 1879.

Holmes, Richard. *The Age of Wonder.* Vintage Books, 2008.

Hoskin, Michael. "Caroline Herschel's Life of 'Mortifications and Disappointments.'" *Journal for the History of Astronomy*, vol. 45, no. 4, 2014, 442–66.

Kawashima, Keiko. "The Evolution of the Gender Question in the Study of Madame Lavoisier." *Historia Scientiarum*, vol. 23, no. 1, 2013, 24–37.

Roberts, Meghan K. "Philosophes Mariés and Espouses Philsophiques: Men of Letters and Marriage in Eighteenth-Century France." *French Historical Studies*, vol. 35, no. 3, 2012, 509–39.

Schiebinger, Londa. *The Mind Has No Sex?: Women in the Origins of Modern Medicine.* Harvard University Press, 1989.

Spradley, Joseph L. "Two Centennials of Star Catalogs Compiled by Women." *The Astronomy Quarterly*, vol. 7, 1990, 177–84.

CHAPTER 5

Messbarger, Rebecca. *The Lady Anatomist: The Life and Work of Anna Morandi Manzolini.* Chicago: University of Chicago Press, 2010.

Messbarger, Rebecca. "Waxing Poetic: Anna Morandi Manzolini's Anatomical Sculptures." *Configurations 9*, no. 1 (2001), 65-97.

Ogilvie, Marilyn and Joy Harvey, eds. *The Biographical Dictionary of Women in Science*, Volume 1. New York: Routledge, 2000.

Park, Katharine. *Secrets of Women: Gender, Generation and the Origins of Human Dissection.* Cambridge, Zone Books, 2010.

San Juan, Rose Marie "The Horror of Touch: Anna Morandi's Wax Models of Hands," *Oxford Art Journal* 34, no. 3 (2011), 433-447.

Schiebinger, Londa. *The Mind Has No Sex? Women in the Origins of Modern Science.* Cambridge: Harvard University Press, 1989.

CHAPTER 6

Knapp, Sandra. "The Plantswoman Who Dressed as a Boy." *Nature*, vol. 470, 2011, 36–37.

Kwadwo Osei-Tutu, John, and Victoria Ellen Smith. "Interpreting West Africa's Forts and Castles." *Shadows of Empire in West Africa: New Perspectives on European Fortifications*, edited by John Kwadwo Osei-Tutu and Victoria Ellen Smith, Palgrave Macmillan, 2018, 1–31.

McEwan, Cheryl. "Gender, Science, and Physical Geography in Nineteenth-Century Britain." *Area*, vol. 30, no. 3, 1998, 215–23.

Merian, Maria Sibylla. *Metamorphosis Insectorum Surinamensium*. Tot Amsterdam, Voor den auteur..., als ook by Gerarde Valck, 1705.

Orr, Mary. "The Stuff of Translation and Independent Female Scientific Authorship: The Case of Taxidermy...,Anon. (1820)." *Journal of Literature and Science*, vol. 8, no. 1, 2015, 27–47.

Ridley, Glynis. The Discovery of Jeanne Baret: *A Story of Science, the High Seas, and the First Woman to Circumnavigate the Globe*. Broadway Books, 2011.

Schiebinger, Londa. *Plants and Empire: Colonial Bioprospecting in the Atlantic World*. Harvard University Press, 2004.

Thompson, Carl. "Women Travellers, Romantic-Era Science and the Banksian Empire." Notes and Records: *The Royal Society Journal of the History of Science*, 2019, 1–25.

Valiant, Sharon. "Maria Sibylla Merian: Recovering an Eighteenth-Century Legend." *Eighteenth-Century Studies*, vol. 26, no. 3, 1993, 467–79.

"The Slavery Connection: Bexley Heritage Trust, 2007-2009." Antislavery Usable Past, *http://www.antislavery.ac.uk/items/show/23.*

CHAPTER 7

Cantor, Geoffrey, et al. *Science in the Nineteenth-Century Periodical*. Cambridge University Press, 2004.

Fenwick Miller, Florence. *Harriet Martineau*. Edited by John H. Ingram, W.H. Allen & Co., 1884.

Larsen, Jordan. "The Evolving Spirit: Morals and Mutualism in Arabella Buckley's Evolutionary Epic." *Notes and Records: Royal Society Journal of the History of Science*, vol. 71, 2017, 385–408.

Leigh, G. Jeffery, and Alan J. Rocke. "Women and Chemistry in Regency England: New Light on the Marcet Circle." *Ambix*, vol. 63, no. 1, 2016, 28–45.

Lightman, Bernard. *Victorian Popularizers of Science: Designing Nature for New Audiences*. University of Chicago Press, 2007.

Lindee, Susan M. "The American Career of Jane Marcet's Conversations on Chemistry, 1806-1853." *Isis*, vol. 82, no. 1, 1991, 8–23.

Sheffield, Suzanne Le-May. *Revealing New Worlds: Three Victorian Women Naturalists*. 1st ed., Routledge, 2013.

Gates, Barbara T. *Kindred Nature: Victorian and Edwardian Women Embrace the Living World*. University of Chicago Press, 1998.

Rauch, Alan. "Parables and Parodies: Margaret Garry's Audience in the Parables from Nature." *Children's Literature*, vol. 25, 1997, 137–52.

CHAPTER 8

Gates, Barbara T. *Kindred Nature: Victorian and Edwardian Women Embrace the Living World*. University of Chicago Press, 1998.

Gianquitto, Tina. "Botanical Smuts and Hermaphrodites: Lydia Becker, Darwin's Botany, and Education Reform." *Isis*, vol. 104, no. 2, June 2013, 250–77.

Kennedy, Meegan. "Discriminating the 'Minuter Beauties of Nature': Botany as Natural Theology in a Victorian Medical School." *Strange Science: Investigating the Limits of Knowledge in the Victorian Age*, University of Michigan Press, 2017, 40–61.

McNeill, Leila. "The Early Feminist Who Used Botany To Teach Kids About Sex." *The Atlantic*, Oct. 2016, *https://www.theatlantic.com/science/archive/2016/10/the-early-feminist-who-used-botany-to-teach-kids-about-sex/503030/#:~:targetText=But%20when%20Elizabeth%20Wolstenholme%20Elmy,as%20a%20sex%2Deducation%20handbook.*

Schiebinger, Londa. *Nature's Body: Gender in the Making of Modern Science*. Rutgers University Press, 1993.

Shteir, Ann B. *Cultivating Women, Cultivating Science: Flora's Daughters and Botany in England 1760-1860*. Johns Hopkins University Press, 1996.

Wolstenholme Elmy, Elizabeth. "Baby Buds." *In Nature's Name: An Anthology of Women's Writing and Illustration, 1780-1930*, edited by Barbara T. Gates, University of Chicago Press, 2002, 484–87.

Wright, Maureen. *Elizabeth Wolstenholme Elmy and the Victorian Feminist Movement: The Biography of an Insurgent Woman*. Manchester University Press, 2011.

CHAPTER 9

"Classified Ad 29," *The New York Times*, September 25, 1861. ProQuest Historical Newspapers.

D'Antonio, Patricia. *American Nursing: A History of Knowledge, Authority, and the Meaning of Work*. Baltimore: The Johns Hopkins University Press, 2010.

Davis, Althea T. *Early Black American Leaders in Nursing: Architects for Integration and Equality*. Jones and Bartlett Publishers and National League for Nursing, 1999.

Afterword

Free, Elizabeth and Liping Bu, "The Origins of Public Health Nursing: The Henry Street Visiting Nurse Service," *American Journal of Public Health* 100, no. 7 (2010), 1206-1207.

Marjorie N. Feld, Marjorie N. Lillian Wald: *A Biography. Chapel Hill.* University of North Carolina Press, 2008.

Nightingale, Florence. *Notes on Nursing: What it is, and what it is not.* New York: D. Appleton and Company, 1860.

Reverby, Susan M. *Ordered to Care: The Dilemma of American Nursing, 1850-1945.* Cambridge: Cambridge University Press, 1987.

Wald, Lillian. *The House on Henry Street.* New York: Henry Holt and Company, 1912.

CHAPTER 10

Josambi, Meera. "Anandibai Joshee: Retrieving a Fragmented Feminist Image." *Economic and Political Weekly* 31, no. 49 (1996): 3189-3197.

Mathes, Valerie Sherer. "Susan La Flesche Picotte, M.D.: Nineteenth-Century Physician and Reformer." *Great Plains Quarterly* 13, no. 3 (1993), 172-186.

Debra Michals, Debra. "Elizabeth Blackwell," National Women's History Museum. Last accessed September 9, 2019. *https://www.womenshistory.org/education-resources/biographies/elizabeth-blackwell.*

McNeill, Leila. "Dr. Anna Fischer-Dückelmann as Naturopath and Physician for Women in Imperial Germany." Master's Thesis, University of Oklahoma, 2014.

—"This 19th Century 'Lady Doctor' Helped Usher Indian Women Into Medicine," *Smithsonian Magazine* August 24, 2017. Last accessed September 9, 2019. *https://www.smithsonianmag.com/science-nature/19th-century-lady-doctor-ushered-indian-women-medicine-180964613/.*

Morantz-Sanchez, Regina Markell. *Sympathy and Science: Women Physicians in American Medicine.* Oxford: Oxford University Press, 1985.

Pripas-Kapit, Sarah Ross. "Educating Women Physicians of the World: International Students of the Women's Medical College of Pennsylvania, 1883-1911. PhD Diss., University of California, Los Angeles, 2015.

Pripas-Kapit, Sarah. "'We Have Lived on Broken Promises': Charles A. Eastman, Susan La Flesche Picotte, and the Politics of American Indian Assimilation during the Progressive Era." *Great Plains Quarterly* 35, no.1 (2015), 51-78.

CHAPTER 11

Brück, Mary. "Slave-Wage Earners." *Women in Early British and Irish Astronomy,* Springer, 2009, 203–20.

Brück, M. T. "Lady Computers at Greenwich in the Early 1890s." *Quarterly Journal of the Royal Astronomical Society,* vol. 36, 1995, 83–95.

Grier, David Alan. When Computers Were Human. Princeton University Press, 2005.

Hoffleit, D. "Women in the History of Variable Star Astronomy." *American Association of Variable Star Observers,* 1993, 1–62.

Kidwell, Peggy Aldrich. "Women Astronomers in Britain, 1780-1930." *Isis,* vol. 75, no. 3, 1984, 534–46.

Klumpke, Dorothea. "The Work of Women in Astronomy." *The Observatory,* 1899.

Mack, Pamela E. "Strategies and Compromises: Women in Astronomy at Harvard College Observatory, 1870-1929." *Journal for the History of Astronomy,* vol. 21, no. 1, 1990, 65–76.

Ogilvie, Marilyn Bailey. "Obligatory Amateurs: Annie Maunder (1868-1947) and British Women Astronomers at the Dawn of Professional Astronomy." *British Journal for the History of Science,* vol. 33, 2000, 67–84.

Rossiter, Margaret W. "'Women's Work' in Science, 1880-1910." *Isis,* vol. 71, no. 3, 1980, 381–98.

Sobel, Dava. *The Glass Universe: How The Ladies of the Harvard Observatory Took the Measure of the Stars.* Viking, 2016.

Stevenson, T. "Making Visible the First Women in Astronomy in Australia: The Measurers and Computers Employed for the Astrographic Catalogue." *Publications of the Astronomical Society of Australia,* vol. 31, 2014, pp. 1–10.

CHAPTER 12

[New York Press] "Raises Rats and Mice." *The Los Angeles Times,* December 26, 1907.

[Brooklyn Eagle] "Woman Runs a Mouse Farm." *The Washington Post,* June 20, 1909.

"Houseworkers and New Apparatus," *Good Housekeeping,* January 1913.

Harriet Gillespie, Harriet. "Labor-Saving Devices Supplant Servants." *Good Housekeeping,* January 1913.

Laurel D. Graham, Laurel D. "Domesticating Efficiency: Lillian Gilbreth's Scientific Management of Homemakers, 1924-1930." *Signs* 24, no. 3 (1999), 633-675.

Leila McNeill, Leila. "The History of Breeding Mice for Science Begins With a Woman in a Barn." Smithsonian. com, March 20, 2018, Last accessed September 18, 2019, *https://smithsonianmag.com/science-nature/history-breeding-mice-science-leads-back-woman-barn-180968441/* .

Miller, Elisa. "In the Name of the Home: Women, Domestic Science, and American Higher Education, 1864-1930." PhD Diss., University of Illinois Urbana-Champaign, 2003.

Rader, Karen. *Making Mice: Standardizing Animals for American Biomedical Research, 1900-1955.* Princeton: Princeton University Press, 2004.

David P. Steensma, David P, et. al., "Abbie Lathrop, the "Mouse Woman of Granby": Rodent Fancier and Accidental Genetics Pioneer," *Mayo Clinic Proceedings* 85, no. 11 (2010, 83.

CHAPTER 13

"Notes," *Nature March* 17, 1921, 88.

"No Gods, No Masters": Margaret Sanger on Birth Control," History Matters, George Mason University. Accessed December 26, 2019. *http://historymatters. gmu.edu/d/5084/.*

Doan, Laura. "Marie Stopes's Wonderful Rhythm Charts: Normalizing the Natural," *Journal of the History of Ideas*, 78, no. 4 (2017), 595-620.

Hajo, Cathy Moran. *Birth Control on Main Street: Organizing Clinics in the United States, 1916-1939.* Champagne, University of Illinois Press, 2010.

—"What Every Woman Should Know: Birth Control Clinics in the United States, 1916-1940." PhD Diss. New York, New York University, 2006.

Hall, Lesley A. "Stopes [married name Roe], Marie Charlotte Carmichael (1880–1958), sexologist and advocate of birth control," *Oxford Dictionary of National Biography*, 2004. Accessed September 24, 2019. *https://www-oxforddnb-com.ezproxy.lib.ou.edu/ view/10.1093/ref:odnb/9780198614128.001.0001/odnb-9780198614128-e-36323.*

Jones, Greta. "Women and eugenics in Britain: The case of Mary Scharlieb, Elizabeth Sloan Chesser, and Stella Browne," *Annals of Science* 52, no. 5 (1995): 481-502.

Katz, Esther. "Sanger, Margaret (14 September 1879–06 September 1966), birth control advocate." American National Biography, 2000. Accessed September 24, 2019. *https://www-anb-org.ezproxy.lib.ou.edu/ view/10.1093/anb/9780198606697.001.0001/anb-9780198606697-e-1500598.*

Kline, Wendy. *Building a Better Race: Gender, Sexuality, and Eugenics from the Turn of the Century to the Baby Boom.* Berkeley: University of California Press, 2005.

Malladi, Lakshmeeramya. "United States v. One Package of Japanese Pessaries (1936)". *Embryo Project Encyclopedia* (2017-05-24). ISSN: 1940-5030 http:// embryo.asu.edu/handle/10776/11516.

Michals, Debra. "Margaret Sanger." National Women's History Museum, 2017. Accessed September 27, 2019. www.womenshistory.org/education-resources/ biographies/margaret-sanger.

Sanger, Margaret. ""The Woman Rebel" and The Fight for Birth Control," [Apr 1916] .Published article. Source: Records of the Joint Legislative Committee to Investigate Seditious Activities, New York State Archives , Margaret Sanger Microfilm, C16:1035. *https://www.nyu.edu/projects/sanger/webedition/app/ documents/show.php?sangerDoc=306320.xml.*

Young, Jennifer. "An Emancipation Proclamation to the Motherhood of America," Lady Science,

November 16, 2017. *https://www.ladyscience.com/ emancipation-proclamation-to-the-motherhood-of-america/ n038.*

CHAPTER 14

Annual Report of the Trustees of the Peabody Museum of American Archaeology and Ethnology. Vol. III, John Wilson and Son, 1887.

Brownman, David L. *Cultural Negotiations: The Role of Women in the Founding of Americanist Archaeology.* University of Nebraska Press, 2013.

Bruchac, Margaret M. Savage Kin: *Indigenous Informants and American Anthropologists.* University of Arizona Press, 2018.

Harrington, M. R. "Ashes Found with Sloth Remains." *The Science News-Letter*, vol. 17, no. 478, June 1930, 365.

—"Man and Beast in Gypsum Cave." The Desert Magazine, vol. 3, no. 6, Apr. 1940, 3–5.

Nuttall, Zelia. "Ancient Mexican Superstitions." *The Journal of American Folklore*, vol. 10, no. 39, 1897, 265–81.

—"New Year of Tropical American Indigenes: The New Year Festival of the Ancient Inhabitants of Tropical America and Its Revival." *Pan American Miscellany*, no. 9, 1928, 2–8.

Ruiz, Carmen. *Insiders and Outsiders in Mexican Archaeology (1890-1930).* The University of Texas at Austin, 2003.

CHAPTER 15

Groves, Leslie R. *Now It Can Be Told: The Story of the Manhattan Project*. Harper & Brothers, 1962.

Higuchi, Toshihiro. "Epistemic Frictions: Radioactive Fallout, Health Risk Assessments, and the Eisenhower Administration's Nuclear-Test Ban Policy, 1954-1958." *International Relations of the Asia-Pacific*, vol. 18, 2018, 99–124.

Howes, Ruth H., and Caroline L. Herzenberg. *Their Day in the Sun: Women of the Manhattan Project*. Temple University Press, 1999.

Kiernan, Denise. *The Girls of Atomic City: The Untold Story of the Women Who Helped Win World War II*. Touchstone, 2013.

Lidofsky, Leon. "Chien-Shiung Wu, 29 May 1912. 16 February 1997." *Proceedings of the American Philosophical Society*, vol. 145, no. 1, 2001, 115–26.

Light, Jennifer S. "When Computers Were Women." *Technology and Culture*, vol. 20, no. 3, 1999, 455–83.

Sumikoh. "A Life Story of Saruhashi Katsuko (1929-2007)." *Contemporary Japanese Feminist Debates at Penn*, 2016, *https://japanfeministdebates. wordpress.com/author/sumikoh/*.

"The S-1 Committee." Atomic Heritage Foundation, 27 Apr. 2017, *https://www.atomicheritage.org/history/s-1-committee*.

CHAPTER 16

Appel, Phyllis. "Dr. Tilly Edinger, 1897-1967, Paleoneurologist." *The Jewish Connection: Profiles of the Famous and Infamous*, Graystone Enterprises LLC, 2013.

Buchholtz, Emily A., and Ernst-August Seyfarth. "The Gospel of the Fossil Brain: Tilly Edinger and the Science of Paleoneurology." *Brain Research Bulletin*, vol. 48, no. 4, 1999, 351–61.

Cavna, Michael. "Emmy Noether Google Doodle: Why Einstein Called Her a 'creative mathematical genius.'" *The Washington Post*, 23 Mar. 2015, *https://www.washingtonpost.com/news/comic-riffs/wp/2015/03/23/emmy-noether-google-doodle-why-einstein-called-her-a-creative-mathematical-genius/*.

Eden, Alp, and Gürol Irzik. "German Mathematician in Exile in Turkey: Richard von Mises, William Prager, Hilda Geiringer, and Their Impact on Turkish Mathematics." *Historia Mathematica*, vol. 39, 2012, 432–59.

Hamilton, Alice. *Exploring the Dangerous Trades: The Autobiography*. Little, Brown and Company, 1943.

Lamb, Evelyn. "Hermann Weyl's Poignant Eulogy for Emmy Noether." Scientific American, Nov. 2016, *https://blogs.scientificamerican.com/roots-of-unity/hermann-weyls-poignant-eulogy-for-emmy-noether/*.

Lang, Henry. "Tilly Edinger's Deafness." *Tilly Edinger: Leben Und Werk Einer Jüdischen Wissenschaftlerin*, edited by Rolf Kohring and Gerald Kreft, vol. 1, Schweizerbart Sche Vlgsb., 2003, 359–72.

Reisman, Arnold. "Hilda Geiringer: A Pioneer of Applied Mathematics and a Woman Ahead of Her Time Was Saved from Fascism by Turkey." *Women in Judaism: A Multidisciplinary Journal*, vol. 4, no. 2, Fall 2007, 1–19.

Rossiter, Margaret W. *Women Scientists in America: Before Affirmative Action, 1940-1972*. Vol. 2, Johns Hopkins University Press, 1995.

Shen, Qinna. "A Refugee Scholar from Nazi Germany: Emmy Noether and Bryn Mawr College." *The Mathematical Intelligencer*, vol. 41, no. 3, 2019, 52–65.

Siegmund-Schultze, Reinhard. Mathematicians Fleeing from Nazi Germany: Individual Fates and Global Impact. Princeton University Press, 2009.

CHAPTER 17

Boris, Eileen. "The Power of Motherhood: Black and White Activist Women Redefine the 'Political.'" *Yale Journal of Law and Feminism*, vol. 2, no. 25, 1989, 25–49.

Terrell, Mary Church. The Progress of Colored Women. National Woman's Suffrage Association, Washington D.C, 1898.

Cole, Rebecca. "First Meeting of the Women's Missionary Society of Philadelphia." *The Women's Era*, vol. 3, no. 4, 1896.

Davis, Jack E. "'Conservation Is Now a Dead Word': Marjory Stoneman Douglas and the Transformation of American Environmentalism." *Environmental History*, vol. 8, no. 1, 2003, 53–76.

Haber, Barbara. "Cooking." *The Reader's Companion to U.S. Women's History*, edited by Wilma Mankiller et al., Houghton Mifflin Company, 1998.

McNeill, Leila. "The First Female Student at MIT Started an All-Women Chemistry Lab and Fought for Food Safety." Smithsonian.com, 2018, *https://www.smithsonianmag.com/science-nature/first-female-student-mit-started-women-chemistry-lab-food-safety-180971056/*.

Merchant, Carolyn. "Women of the Progressive Conservation Movement: 1900-1916." *Environmental Review*, vol. 8, no. 1, 1984, 57–85.

Musil, Robert K. *Rachel Carson and Her Sisters, Extraordinary Women Who Have Shaped America's Environment.* Rutgers University Press, 2014.

Peine, Mary Anne. *Women for the Wild: Douglas, Edge, Murie and the American Conservation Movement.* University of Montana, Apr. 2002.

Richards, Ellen Henrietta. *The Art of Right Living.* Whitcomb & Barrows, 1904.

Stoneman Douglas, Marjory. *The Everglades: River of Grass.* Rinehart & Company, Inc., 1947.

Unger, Nancy C. *Beyond Nature's Housekeepers: American Women in Environmental History.* Oxford University Press, 2012.

CHAPTER 18

"The Significant of the 'Doll Test,'" NAACP Legal Defense Fund, Accessed November 29, 2019, *https://www.naacpldf.org/ldf-celebrates-60th-anniversary-brown-v-board-education/significance-doll-test/*.

Burke, May Edwin Mann. "The Contributions of Flemmie Pansy Kittrell to Education Through Her Doctrines on Home Economics." PhD Diss University of Maryland, 1988.

Clark, Mamie Phipps. "Interview of Dr. Mamie Clark by Ed Edwin, May 25, 1976." Interview by Ed Edwin in *The Reminiscences of Mamie Clark.* Alexandria: Alexander Street Press, 2003.

Horrocks, Allison Beth. "Good Will Ambassador with a Cookbook: Flemmie Kittrell and the International Politics of Home Economics." PhD Diss, University of Connecticut, 2016.

Karera, Axelle (2010) and Alexandra Rutherford (2017). "Profile of Mamie Phipps Clark," in *Psychology's Feminist Voices Multimedia Internet Archive*, ed. A. Rutherford. Accessed December 23, 2019, *http://www.feministvoices.com/mamie-phipps-clark/*.

Malcom, Shirley Mahaley, et. al. "The Double Bind: The Price of Being a Minority Woman in Science." *Conference Proceedings, American Association for the Advancement of Science* (76-R-3), 1976.

Scriven, Olivia A. "The Politics of Particularism: HBCUs, Spelman College, and the Struggle to Educate Black Women in Science, 1960-1997." PhD Diss, Georgia Institute of Technology, 2006.

CHAPTER 19

Archival material from the Kennedy Space Center Files, Office of Manned Space Flight, Public Affairs Office, National Archives and Records Administration, Atlanta, GA.

"NASA Johnson Space Center Oral History Project Biographical Data Sheet, Dee O'Hara." National Aeronautics and Space Administration, Johnson Space Center Oral History Project. *https://historycollection.jsc.nasa.gov/JSCHistoryPortal/history/oral_histories/OHaraDB/OHaraDB_Bio.pdf.*

Dee O'Hara, NASA Johnson Space Center Oral History Project Oral History Transcript, Dee O'Hara," interviewed by Rebecca Wright, Mountain View California, April 23, 2002. *https://historycollection.jsc.nasa.gov/JSCHistoryPortal/history/oral_histories/OHaraDB/OHaraDB_4-23-02.pdf.*

NASA Biographical Data, "Mae C. Jemison," National Aeronautics and Space Administration (nd): *https://www.nasa.gov/sites/default/files/atoms/files/jemison_mae.pdf.*

NASA Biographical Data, "Chiaki Muka," National Aeronautics and Space Administration (nd): *https://www.nasa.gov/sites/default/files/atoms/files/mukai.pdf.*

"Valentina Tereshkova," Smithsonian National Air and Space Museum, Accessed October 24, 2019. *https://airandspace.si.edu/people/historical-figure/valentina-tereshkova.*

Clements, Barbara Evans. "Tereshkova, Valentina," in *The Oxford Encyclopedia of Women in World History*, ed. Bonnie G. Smith. Oxford: Oxford University Press, 2008.

Foster, Amy E. *Integrating Women into the Astronaut Corps: Politics and Logistics at NASA, 1972-2004.* Baltimore: The Johns Hopkins University Press, 2011.

Reser, Anna. "The Lost Stories of NASA's 'Pink-Collar' Workforce, The Atlantic February 15, 2017. *https://www.theatlantic.com/science/archive/2017/02/ursula-vils-nasa/516468/.*

Sanders, Matthew. "The Woman Who Got Real Food to Space," Smithsonian National Air and Space Museum, April 9, 2018. *https://airandspace.si.edu/stories/editorial/woman-who-got-real-food-space.*

Siddiqi, Asif A. *Sputnik and the Soviet Space Challenge.* Gainesville, University Press of Florida, 2003.

CHAPTER 20

Capshew, James H., and Alejandra C. Laszlo. "'We Would Not Take No for an Answer': Women Psychologists and Gender Politics During World War II." *Journal of Social Issues*, vol. 42, no. 1, 1986, 157–80.

Held, Lisa. "Leta Hollingworth." *Psychology's Feminist Voices*, 2010, http://www.feministvoices.com/leta-hollingworth/.

Johnson, Amy, and Elizabeth Johnston. "Unfamiliar Feminisms: Revisiting the National Council of Women Psychologists." *Psychology of Women Quarterly*, vol. 34, 2010, 311–27.

Lowie, Robert H., and Leta Stetter Hollingworth. "Science and Feminism." *The Scientific Monthly*, vol. 3, no. 3, 1916, 277–84.

Mednick, Martha T., and Laura L. Urbanski. "The Origins and Activities of APA's Division of the Psychology of Women." *Psychology of Women Quarterly*, vol. 15, 1991, 651–63.

Morawski, Jill G., and Gail Agronick. "A Restive Legacy: The History of Feminist Work in Experimental and Cognitive Psychology." *Psychology of Women Quarterly*, vol. 15, 1991, 567–79.

Rutherford, Alexandra, and Leeat Granek. "Emergence and Development of the Psychology of Women." *Handbook of Gender Research in Psychology*, edited by J.C. Chrisler and D.R. McCreary, Springer Science+Business Media, LLC, 2010.

Rutherford, Alexandra, et al. "Responsible Opposition, Disruptive Voices: Science, Social Change, and the History of Feminist Psychology." *Psychology of Women Quarterly*, vol. 34, 2010, 460–73.

Shields, Stephanie A. "Ms. Pilgrim's Progress: The Contributions of Leta Stetter Hollingworth to the Psychology of Women." *American Psychologist*, vol. 30, no. 8, 1975, 852–57.

Stetter Hollingworth, Leta. "Variability as Related to Sex Differences in Achievement: A Critique." *American Journal of Sociology*, vol. 19, no. 4, 1914, 510–30.

Thorndike, Edward Lee. *Educational Psychology: Mental Work and Fatigue and Individual Differences and Their Causes.* Teacher's college, Columbia University, 1921.

Weisstein, Naomi. "'How Can a Little Girl like You Teach a Great Big Class of Men?' The Chairman Said, and Other Adventures of a Woman in Science." *Working It Out: 23 Women Writers, Artists, Scientists, and Scholars Talk About Their Lives and Work*, edited by Sara Ruddick and Pamela Daniels, Pantheon Books, 1977, 241–50.

Weisstein, Naomi. "Psychology Constructs the Female; or, The Fantasy Life of the Male Psychologist (with Some Attention to the Fantasies of His Friends, the Male Biologist and the Male Anthropologist)." *Feminism & Psychology*, vol. 3, no. 2, 1993, 195–210.

CHAPTER 21

"Coeducation: History of Women at Princeton University," Princeton University, Accessed November 29, 2019, *https://libguides.princeton.edu/c.php?g=84581&p=543232*.

"A History of Palomar Observatory, Caltech, Accessed November 29, 2019, http://www.astro.caltech.edu/palomar/about/history.html#55

"The Nobel Prize in Physics 1974," Accessed December 23, 2019. *https://www.nobelprize.org/prizes/physics/1974/summary/*

"The Nobel Medal for Physics and Chemistry," The Nobel Prize. Accessed November 29, 2019, *https://www.nobelprize.org/prizes/facts/the-nobel-medal-for-physics-and-chemistry-2*.

Burnell, Jocelyn Bell. "Petit Four," *Annals of the New York Academy of Sciences* 302, no. 1 (1977), 685-689.

Burnell, "So Few Pulsars, So Few Females," Science 304, no. 5670 (2004), 489

Lambourne, Rober. "Interview with Jocelyn Bell Burnell," *Physics Education* 31, no. 183 (1996), 183-186

Larsen, Kristine. "Reminiscences on the Career of Martha Stahr Carpenter" Between a Rock and (Several) Hard Places," *Journal of the American Association of Variable Star Observers* 40 (2012), 51-64

Rubin, Vera C. "An Interesting Voyage," *The Annual Review of Astronomy and Astrophysics* 49 (2011), 1-28.

Bibliography

First published in 2021 by Frances Lincoln Publishing
an imprint of The Quarto Group.
The Old Brewery, 6 Blundell Street
London, N7 9BH,
United Kingdom
T (0)20 7700 6700
www.QuartoKnows.com

A catalogue record for this book is available from the British Library.

ISBN   978-0-7112-4897-7
Ebook ISBN   978-0-7112-4898-4

10 9 8 7 6 5 4 3 2 1

Publisher: Philip Cooper
Editorial Director: Jennifer Barr
Picture Researcher: Bella Skertchly
Designer: Isabel Eeles

Printed in China